W9-DFD-591

BRUMBACK LIBRARY
3 3045 00112 2817

THE CHINESE AMERICAN
FAMILY ALBUM

THE CHINESE AMERICAN FAMILY ALBUM

DOROTHY AND THOMAS HOOBLER

Introduction by Bette Bao Lord

OXFORD UNIVERSITY PRESS • NEW YORK • OXFORD

Authors' Note

Scholars have long struggled with the problem of romanizing the Chinese language—writing Chinese with the 26 letters of the English alphabet. In the 19th century, two British scholars, Sir Thomas Wade and H. A. Giles, developed a system based on the pronunciations of Mandarin (northern) Chinese. Most English-language books about China followed the Wade-Giles system or a modified version of it until 1979. That year, the Chinese press agency decided to employ a new system, called pinyin. The *New York Times* adopted pinyin spellings and since then, so have most book publishers.

As we gathered the selections for this book, we found that the Chinese Americans who wrote them used many different kinds of romanizations. In all cases, we have let the writers' original spellings stand because we wanted to allow Chinese Americans to speak for themselves.

In the introductions that we wrote, we have generally used pinyin spellings. However, to avoid confusion, we used the Chinese Americans' own spellings to introduce some chapters or selections. This is not a solution that will satisfy everyone, but we felt it was the clearest way to give our readers an insight into Chinese American history and life.

Cover: *Lai Ngan (center left) and Lee Kwong (center right) with their children in Nogales, Arizona, around 1905.*

Frontispiece: *Mr. and Mrs. Fong Wan and their children in California around 1910.*

Contents page: *A 19th-century miner in Arizona.*

AS

Oxford University Press

Oxford New York Toronto
Delhi Bombay Calcutta Madras Karachi
Kuala Lumpur Singapore Hong Kong Tokyo
Nairobi Dar es Salaam Cape Town
Melbourne Auckland Madrid

and associated companies in
Berlin Ibadan

Design: Sandy Kaufman
Consultant: Franklin Ng, Professor of Anthropology, California State University at Fresno

Published by Oxford University Press, Inc.,
200 Madison Avenue, New York, New York 10016

Library of Congress Cataloging-in-Publication Data

Hoobler, Dorothy.
The Chinese American family album / Dorothy and Thomas Hoobler.
p. cm. — (American family albums)
Includes bibliographical references and index.
1. Chinese Americans—History —Juvenile literature. [1. Chinese Americans—History.]
I. Hoobler, Thomas. II. Title. III. Series.
E184.C5H66 1994 93-11873
CIP
973'.04951—dc20 AC

ISBN 0-19-508130-7 (lib. ed.); ISBN 0-19-509123-X (trade ed.); ISBN 0-19-509125-6 (series, lib. ed.)

3 5 7 9 8 6 4 2

Printed in the United States of America
on acid-free paper

22.95

CONTENTS

Bette Bao in front of her house in Brooklyn in 1947.

Bette Bao (center) with her parents (in the back row) in China.

Shanghai-born Bette Bao Lord is the author of two novels, Spring Moon, *which was nominated for the American Book Award, and* Eighth Moon, *about her youngest sister's life in China. Mrs. Lord is also the author of a children's book,* In the Year of the Boar and Jackie Robinson, *which is based on her early years in the United States. Her moving and richly textured account of contemporary China,* Legacies, *conveys the experiences of members of her own family still living in China as well as those of a broad spectrum of friends throughout the country. Mrs. Lord was co-producer of the People's Art Theatre, Beijing, production of* The Caine Mutiny, *directed by Charlton Heston, and a consultant to CBS News for its coverage of the Tienanmen Square events in 1989. She has been honored with the American Women for International Understanding award, the National Committee for U.S.-China Relations award, and the Woman of the Year award from the Chinatown (New York) Planning Council. She is married to Winston Lord, Assistant Secretary of State for East Asian and Pacific Affairs and former U.S. ambassador to China, and they have two children.*

Bette Bao Lord signing copies of Spring Moon *in China in 1988.*

Bette Bao Lord with her husband, Winston Lord (far left), and their son and her parents.

INTRODUCTION

by Bette Bao Lord

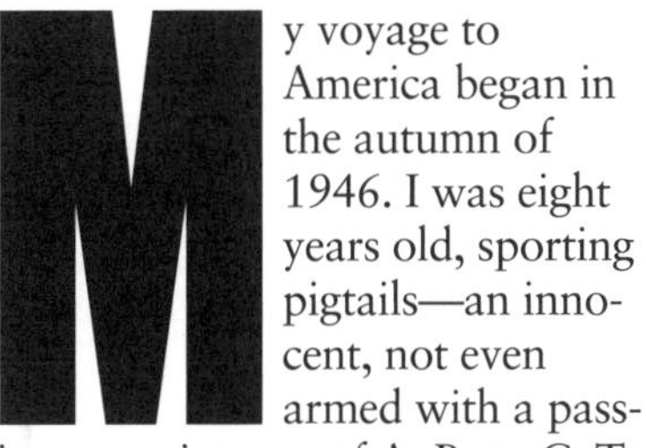y voyage to America began in the autumn of 1946. I was eight years old, sporting pigtails—an innocent, not even armed with a passing acquaintance of A, B, or C. To my chagrin, the ocean was not the vast jade lagoon that I had always envisioned but about as pacific as a fierce dragon with chilies up its snout. And so I bravely cowered in my bunk battling to keep down what I assumed was an authentic American delicacy—spaghetti with meatballs.

Only yesterday, resting my chin on the rails of the S.S. *Marylinx,* I peered into the mist for *Mei Guo,* beautiful country. It refused to appear. Then, within a blink, there was the golden gate, more like the portals to heaven than the arches of a man-made bridge.

I arrived in Brooklyn, New York, on a Sunday. On Monday I was enrolled at P.S. 8. By putting up 10 fingers, I found myself sentenced to the fifth grade. It was a terrible mistake. By American reckoning, I had just turned eight. And so I was the shortest student by a head or two in class. In retrospect, I suppose that everyone just supposed that Chinese were supposed to be small.

Only yesterday, holding my hand over my heart, I joined schoolmates to stare at the Stars and Stripes and say along: "I pledge a lesson to the frog of the United States of America. And to the wee puppet for witches' hands. One Asian, in the vestibule, with little tea and just rice for all."

Only yesterday, rounding third base in galoshes, I swallowed a barrelful of tears wondering what wrong I had committed to anger my teammates so. Why were they all madly screaming at me to go home, go home?

Only yesterday, parroting the patter on our Philco radio, I mastered a few mouthfuls of syllables and immediately my teacher began eliciting my opinions. Not only was I stupefied by the challenge of puzzling together "Pepsi-Cola hits the spot," "I'd walk a mile for a Camel," or "Hi-ho Silver," into a coherent thought, but I was amazed by the fact that an exalted teacher would solicit the opinion of a lowly student. Teachers in China never did that.

Eventually, I came to realize that the merits of one's opinions were not the determining goal of the exercise. The goal was to nurture a civil society where everyone is free to speak. Today, when political correctness threatens the rigor of our intellectual debates, how I value this aspect of my early education! To me, the cacophony of puddingheads spewing their views is preferable to the clarion call of even the greatest emperor.

Only yesterday, standing still a head or two short at graduation, I felt as tall as the Statue of Liberty as I recited Walt Whitman: "I hear America singing, the varied carols I hear... Each singing what belongs to him or her and to none else."

Thus I have never forgotten that one need not lose one's native culture in order to become an American. On the contrary, this individual feels doubly blessed. For to me, Americans—though as different as sisters and brothers are—belong to the same family. For to me, America is a road cleared by the footfalls of millions of immigrants and paved with something far more precious than gold—grit and hope.

A middle-class couple in 19th-century China.

CHAPTER ONE

THE MIDDLE KINGDOM

In A.D. 499, a Buddhist monk named Hui-shen returned to China after a long journey. He told of a visit he had made across the sea to a land called Fu-Sang, which lay far to the east. Modern scholars have speculated that Fu-Sang might actually have been America.

A tantalizing clue to this possibility is the Monterey cypress tree, which grows on the coast of California and nowhere else in the Western Hemisphere. A naturalist took a branch of one such tree to the Panama-Pacific Exposition in 1915. When she showed it to a visiting Buddhist monk, he recognized that it was similar to a kind of cypress that grows in China. Could Hui-shen have planted the seeds of this tree in California? If so, he was the first Chinese visitor to today's United States.

Fourteen hundred years later, the first Chinese immigrants arrived in the United States, fleeing famine and revolution. In the 19th century, Chung Kuo (the Middle Kingdom), as the Chinese called their homeland, was a country in decline. An alien dynasty of Manchu emperors had ruled the Chinese Empire since 1644. Most Chinese resented the Manchus, and the government's unpopularity rose when it proved unable to deal with the threat of European domination.

Europeans sought to trade their goods—such as opium—with the Chinese Empire, but they met resistance. The Manchu dynasty restricted foreigners to the port of Canton, in Kwangtung Province on the southeast coast. But after British warships demonstrated their superior power in the First Opium War (1839–42), China was forced to open more ports to foreign trade. As a further penalty for its defeat, China ceded to Great Britain the island of Hong Kong.

The intrusion of the West came at a time when China was experiencing a rapid rise in population. In 1700, China's population was about 150 million. It had swelled to 400 million by 1850. This put enormous strains on the farmers to produce enough food to support the people. A small number of landlords controlled much of the farmland on which the majority of Chinese toiled in poverty.

Suffering and famine led to a peasant uprising, the Taiping Rebellion (1850–64), which shook the empire to its roots. More than 20 million people died in the fighting, and the disruption caused more economic hardship. Some Chinese began to flee their homeland to seek opportunity elsewhere.

The vast majority of the early immigrants to the United States came from Kwangtung Province, in the southern part of China. Canton, the provincial capital, and seven districts in the Pearl River delta region of Kwangtung provided more than 95 percent of the 19th-century immigrants. Officially, the Chinese Empire forbade its citizens to leave the country. Even so, Chinese had been emigrating since ancient times. There were many Chinese communities throughout Southeast Asia. Because the British-held island of Hong Kong was off the southern coast of Kwangtung, it was relatively easy for people to go there on small boats and book passage to the United States.

Two distinct groups of people, the Punti and the Hakka, lived in the villages of Kwangtung. The Punti were the original inhabitants of the area. The Hakka were descendants of northern Chinese who had spread to the region after China expanded to the south. A third group, the Tanka, lived in boats along the coast, where they had practiced their trade as fishermen throughout Chinese history. Each of these groups spoke a different dialect and they were often rivals--both at home and overseas. Fighting between the Punti and the Hakka devastated many Chinese villages and spurred the flight overseas. The *Hsin-ning hsien-chih* (Gazetteer of the Hsin-ning District) described hardships in the Pearl River delta region caused by fighting between Punti and Hakka: "The fields in the four directions were choked with weeds. Small families found it difficult to make a living and often drowned their girl babies because of the impossibility of looking after them."

These earliest immigrants were accustomed to a tropical climate. In the delta region of Kwangtung Province, long, hot summers from April to October are followed by cool, dry winters and two months of muggy weather. Seasonal monsoons bring abundant rainfall, making rice the primary crop. In the wintertime, the farmers raised ducks in the rice paddies. Other crops include sugarcane, vegetables, litchi nuts, bananas, mangoes, plums, and oranges. All these foods became ingredients in Chinese American cooking.

In addition, the farmers of Kwangtung planted mulberry trees so that silkworms could feed on the leaves. Chinese women learned the delicate task of unwinding the silkworm cocoons to produce China's most prized cloth.

The traditional Chinese way of life was guided by the teachings of Confucius, a philosopher who lived 2,500 years ago. All Chinese government officials had to pass a grueling series of tests on Confucian literature to obtain their posts. These officials, sometimes called mandarins, were at the very top of Chinese society.

Confucian values included a reverence for the past as an example for the future, respect for elders, and worship of ancestors. Ancestor worship linked families closely and tied the Chinese to their native villages. Every Chinese wanted to be buried in his home village, where he knew that generations of descendants would honor his memory.

A view of the city of Canton, the capital of Kwangtung Province, around 1864. Before 1842, Canton was the only port open to foreign goods. For centuries, it also served as the departure point for Chinese leaving the country.

Confucianism emphasized the importance of family. The Chinese ideal was a large family with several generations living under one roof. This custom was most common for the wealthy, but it was not unusual for married couples of any social class to live with their children in the home of the husband's parents.

There was no higher Confucian ideal than filial piety, the respect and reverence for parents. "Filial piety," wrote one Chinese consul to the United States, "is a cardinal virtue my parents have brought over from China.... A Chinese child, no matter where he lives, is brought up to recognize that he cannot shame his parents.... Before a Chinese child makes a move, he stops to think what the reaction of his parents will be."

Confucian society was male-dominated. The female's place was in the home, and a wife owed obedience both to her husband and to his parents. Marriages were commonly arranged by the parents, and young women had no say in the decision. Men could take concubines, or secondary wives, although this was the case only with the wealthy. The custom of binding the feet of young girls—a painful process that ensured tiny "lily feet"—often made women virtual cripples who seldom left their home. Foot binding was not so widespread among farm families, where women were needed to work in the fields.

The Chinese found truth in several religions, and saw no conflict in practicing all of them. Over time, Buddhism and Taoism blended with the teachings of Confucius. Also, a folk religion provided gods for all occasions. The Chinese view was that heaven, earth, and the underworld had spirits whose actions could affect humankind. A Chinese offered prayers to whichever gods seemed appropriate for the occasion.

The California gold rush of 1849–50 started the first great wave of Chinese immigration to the United States. News of the discovery of gold on the Sacramento River in 1848 excited people even in remote Chinese villages. Stories about Gum Shan (the Golden Mountain) created dreams of finding great wealth. It was rumored that nuggets of gold lay on the ground, just waiting for someone to pick them up.

Even earlier, Chinese had gone to Hawaii, which was at that time an independent kingdom. Hawaii had another mountain of riches that attracted the Chinese—sandalwood, which was greatly prized in China, where it was used to create beautiful furniture.

Chinese immigrants continued to flock to the United States after the gold rush ended. At that time, the United States was a growing country that needed workers. Even menial jobs paid high wages, by Chinese standards. Some Chinese returned home with the wealth they had earned, prompting others to try their luck in the new country.

In the 19th century, the great majority of these Chinese were sojourners. They planned to make money and return to China, rather than putting down roots in the United States. Most of them were farmers who left their wives and families behind. The vast majority were male because in the Confucian system, women were supposed to stay home and take care of their children and in-laws.

There was also a small number of merchants among the immigrants. In China, merchants had low status, but in America they would become the leaders of the Chinese community. Indeed, many former farmers became merchants in the New World.

In 1882, the United States prohibited Chinese immigration. The ban lasted for 60 years. Some resourceful Chinese found ways to get around it, but large-scale immigration did not resume until after World War II.

Treasury Street in Canton, around 1860. Tea and silk were China's most important export goods, but Europeans and Americans also desired the black-lacquered furniture, palm-leaf fans, firecrackers, ivory carvings, toys, and tableware made by artisans in Chinese shops and factories.

Political upheaval in China caused an influx of new immigrants. In 1949, China's long civil war ended, and the communist victors established the People's Republic of China. The partisans of the losing side, the Nationalists, set up their own government on the offshore island of Taiwan. In the years since then, refugees from the People's Republic have made their way to the United States in search of freedom.

When the United States and the People's Republic of China established diplomatic relations in 1979, it became easier for Chinese to emigrate from the mainland. A new communist leadership under Deng Xiaoping allowed greater freedoms. However, a budding democracy movement led by Chinese students was abruptly crushed in June 1989 when tanks moved against demonstrators in Tienanmen Square in the Chinese capital of Beijing. Political refugees have continued to arrive in the United States since that time.

Chinese living in Taiwan have also immigrated to the United States for educational and economic opportunities. Another group of today's Chinese immigrants comes from the British colony of Hong Kong, which will revert to the control of mainland China in 1997. Fearing an uncertain future under communist rule, many of Hong Kong's residents have obtained visas to live in the United States.

During the past two centuries of immigration, the Chinese have brought their great skills and talents to this country. The Confucian ethic, which includes respect for hard work and education, has made them valuable citizens. Chinese helped to build the great railroads that linked the two coasts of the United States. The skill of Chinese farmers helped to make California one of the richest farming areas of our country. In modern times, Chinese American scientists, business leaders, and artists have enriched the life of their new country. Through them, the ancient culture of China has become a part of America.

IMAGES OF THE HOMELAND

Most of the early Chinese immigrants to the United States came from Kwangtung Province in southeast China, as shown on this map of present-day China. Today, many come from other regions, including Hong Kong and the island nation of Taiwan.

The Chinese who came to America retained strong memories of their families and homes in China. In 1931, 78-year-old Huie Kin set down his recollections of his boyhood in southern Kwangtung Province.

Father and mother had their hands full bringing us up with what little they could raise on their small farm. Our home had two large rooms. It was the common practice to keep chickens and pigs in the courtyard; but the family cow, because of its importance, shared the rooms with the family. I remember that we had a cow with velvety brown fur and short, curved horns. Some of the neighbors had water buffaloes with ugly black bristles and unusually long and heavy horns. These were not so pleasant as roommates. My father and I and the brown cow had one of the rooms, with the kitchen stove in one corner; and mother had the other room. Near the door was the *Men-kong,* the deity who was supposed to keep evil out of the house; above the cattle stall was the animal deity; while over the stove presided the kitchen god. These figures were painted on red paper pasted on the wall.

In 1906, Lee Chew, a successful merchant in New York's Chinatown, recalled his boyhood in a village near Canton in the 1860s.

When I was a baby I was kept in our house all the time with my mother, but when I was a boy of seven I had to sleep at nights with other boys of the village—about thirty of them in one house. The girls are separated the same way.... In spite of the fact that any man may correct them for a fault, Chinese boys have good times and plenty of play. We played games like tag, and other games like shinny and a sort of football called *yin.* We had dogs to play with—plenty of dogs and good dogs—that understand Chinese as well as American dogs understand American language. We hunted with them, and we also went fishing and had as good a time as American boys, perhaps better, as we were almost always together in our house, which was a sort of boys' club house, so we had many playmates.... But all our play outdoors was in the daylight, because there were many graveyards about and after dark, so it was said, black ghosts with flaming mouths and eyes and long claws and teeth would come from these and tear to pieces and devour any one whom they might meet.

It was not all play for us boys, however. We had to go to school, where we learned to read and write and to recite the

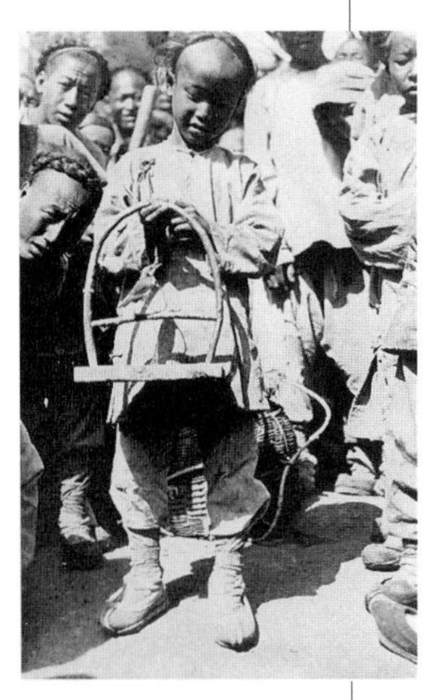

A Chinese boy shows off his trained bird (around 1900). Acrobats, jugglers, magicians, and storytellers often performed in the streets of Chinese cities.

precepts of Kong-foo-tsze [Confucius] and the other Sages, and stories about the great Emperors of China, who ruled with the wisdom of gods and gave to the whole world the light of high civilization and the culture of our literature, which is the admiration of all nations.

A few of the earliest immigrants came from prosperous families. Wong Bing Woo, whose father was an exporter of shark fins, married a scholar who brought her to the United States after he had established a practice as an herb physician. Wong Bing Woo's daughter, a journalist, describes her mother's childhood in China around 1880.

The family compound consisted of 10 grey brick buildings with curving Chinese roofs, with a family unit occupying each house. There were guest houses for visitors, separate ones for men and women. An eight-foot high wall of grey brick surrounded the compound. A barred gate, attended by servants, opened into a courtyard. This elaborate layout was called "Gim Sing Tong," meaning "Built With Money Earned by Industry," and covered ground equal to an average American city block.

Parklike gardens graced the estate. There were trees of all kinds: kumquat, apricot, lemon, lichee nut, loquat, and tangerine. Flowers grew profusely. One, an orange flower called *dan gway,* was as tall as the buildings and bloomed in September. Its fragrance permeated the area even beyond the . . . gate.

At home I obey my parents.
At school I obey my teachers.
In school
Respect the teacher
Love your classmates.

—From a Chinese primer

Nineteenth-century upper-class Chinese families, like this one, had a luxurious life. It was the custom for both men and women to shave the hair above their foreheads. Women wore white makeup and painted their eyebrows and lips.

May you be blessed with many sons!

—Common Chinese greeting to a friend, the usual way of wishing him good luck

Each family member had a servant whose sole duty was to attend to his or her needs. Mama had a nurse (*goo mah*) and later on a maid (*mui jay*).... Mama's leisurely day began at ten, when she rose and breakfasted on *jook* (rice gruel).... The little girls played house with toy dishes and cooking utensils, but they had no dolls. There were games like blind man's bluff, hide and seek, and improvised shows where the children would dress up in their parents' clothes. Adults told them stories, mostly tales that dealt with ethical conduct and the necessity of filial devotion to parents....

Between the ages of 8 and 10, Mama learned to sew and embroider.... In those days, it was thought education for girls was unnecessary, but Mama's father was progressive. He wanted his daughters to learn; so Mama was taught to read by a tutor at the age of eight, and to hold her brush and write characters. Later, she learned a little history and geography as well as some arithmetic, including use of the abacus. However, at age 12 her tutoring was stopped because her tutor was a man....

The Wong girls were taught early how to beautify themselves. Mama was only five when her ears were pierced for earrings. On special holidays, she was allowed to use powder, rouge, and to color her lips with red-coated paper creased in accordion folds.... Probably the most traumatic event of Mama's childhood was the binding of her feet at age six. It would have been done earlier, but her mother's death required that three years pass before the binding. Aunt Beautiful Pearl bound the feet with cotton cloths. It was a tortuous process, and Mama cried with pain for months. Her feet were not unbound until after [the birth of her own child]. Mama was extremely sensitive on the subject of her bound feet. She was ashamed of them and didn't want to talk about the Chinese custom of lily feet. None of us [her children] ever saw her bare feet; she always wore white sox over them. Her bound feet were a lifelong burden, restricting her activities in every way.

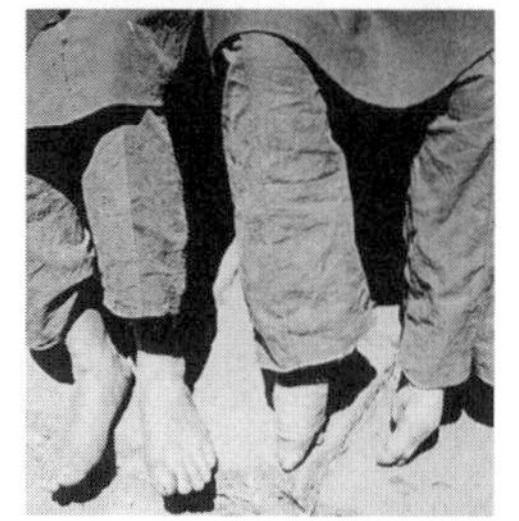

Because tiny "lily feet" were regarded as a mark of feminine beauty, young girls had their feet tightly bound to stunt their growth. A painful process that made women virtual cripples, foot binding was usually an upper-class practice. Farm families could not afford to lose the work of women in the fields. The practice has been banned in modern China.

Chung Kun-ai was born in China in 1865. Fourteen years later he immigrated to Hawaii, where he lived for the rest of his long life. In 1958, he recalled fondly his boyhood in China.

Life in a Chinese village was very interesting and attractive to us youngsters, especially our lunar New Year celebrations. That was the one occasion during the whole year that the villagers took off a whole two weeks to enjoy themselves. For weeks before the New Year's, preparations had been going on. The wine vats had been filled with rice and other ingredients and set aside to allow the grain to ferment. Every year our family stored away enough jugs of yellow rice wine to replenish those consumed....

Glutinous rice had to be ground into rice-flour for New Year pastries. The house had to be cleaned and the kitchen god had to be escorted to heaven a week before New Year's. Everyone had to have new and colorful gowns and other garments,

One of the tasks of women on farms was to grind rice into flour, as the mother of this family is doing. The man in the photo may not be her husband, but her father-in-law. In the Chinese extended family, a married woman was expected to serve her husband's parents until their death.

and these had to be sewed and embroidered by hand. Members of the family who had left the village to earn a living elsewhere also tried their best to make this annual pilgrimage back to the ancestral home, in time to participate in the family gathering on New Year's Eve. Each family closed the old year with as sumptuous a feast as it could afford. The dining and wining went on till midnight, when firecrackers were kept burning and the proper obeisance paid to all the gods that ruled the Chinese universe....

We youngsters could do nothing wrong during this season, for New Year's was the only time during the whole year when it was inauspicious for anyone to scold or nag.... And so we did as we well pleased and none dared defy the gods by chastising us when we were naughty. If only the whole year could be one New Year's!

Our elders...paraded through the villages with their dragons or lions, prancing and dancing as firecrackers were thrown at their feet. Shopkeepers and the wealthy in the village would show their appreciations and also wish themselves greater wealth by tying a gift of money in red paper with a spring green of vegetable leaf, just high enough above the gate to call for special effort by the dancers to collect their "fee." The Chinese words for green vegetables (*sung choi*) and those for growing riches (*sung choi*) are homonyms, and what Chinese does not want to grow richer?...The front porch of our home was paved with granite, and we used to gather there in the eve-

Growing rice, the staple food of China, was an arduous task. Each seedling had to be transplanted by hand into the flooded fields, where it would grow to maturity. Here, a water buffalo is harnessed to a machine that pumps water into the rice fields.

The Tanka fished the waters along the south China coast for centuries. Tanka families lived on the boats that went to sea each day and returned filled with fish that they sold on the wharves of cities like Canton.

nings after supper to listen to our elders. Someone would start off with a story from the Three Kingdoms period, perhaps one of the many stories of K'ung Ming, the sagest tactician in that period of Chinese history, of how he outwitted the enemy.... And the story-teller, whenever the action became exciting, would burst into lines of poetry that said so much in so few words, often chanting the words in sing-song fashion that appealed to us. And that night, before we could fall asleep, we must repeat those lines of poetry until they too became a part of our being. We had become heirs to our great literary tradition.

Some childhoods were not so happy. In 1936, an old woman known as Mrs. Teng told her life story to an interviewer. Mrs. Teng pushed back her black hair with her gnarled hands, worn from years of work. She recalled how she came to Hawaii at the age of nine.

Lucky come Hawaii? Sure, lucky, come Hawaii. Before I come to Hawaii I suffer much. Only two kinds of people in China, the too poor and the too rich. I never can forget my days in China....

In a small crowded village, a few miles from Hongkong, fifty-four years ago I was born. There were four in our family, my mother, my father, my sister and me. We lived in a two room house.... How can we live on six baskets of rice which were paid twice a year for my father's duty as a night watchman?...

Sometimes we went hungry for days. My mother and me would go over the harvested rice fields of the peasants to pick the grains they dropped. Once in a while my mother would go near a big pile of grain and take a handful. She would then sit on them until the working men went home....

Father was suffering from dysentery so my mother went out to look for herbs. My father told me to take the baby out to play and not to come back until late. Being always afraid of him I gladly took the baby out. We were three houses away watching a man kill a chicken. Pretty soon a man came to call me to go home for my father is dead.... I took one look at my father dangling from the ceiling and started to run to where I don't know....

My mother called me to her and put me on her lap. "Do you want me to remarry or will you be a good girl and go to stay with a certain lady," she said. I told her that I do not want her to remarry but I will go with the lady so that she will have money to pay for my father's coffin.... I leaned my head against her breast and...I knew that was the last time I would be so near to her.

I heard my mother tell this go-between lady that she wants me put in the hands of a lady or man who would come to Hawaii because she has heard Hawaii is a land of good fortune.

An Wang, the founder of Wang Laboratories, grew up in China in the 1920s, coming to the United States in 1945 as a student. He recalled his roots in his autobiography, Lessons.

Like many Chinese families, we had a written history which would be updated every couple of generations.... These books gave our families a sense of continuity and permanence that I don't see in the more mobile West. It was only a few years ago [after coming to the United States] that I acquired a copy of this book, but I recall my father pointing it out on a bookshelf when I was a child. It claims to be accurate for twenty-three generations, back to the time of the Mongol invasion and Marco Polo's trip to China. With less certainty, the history goes back another twenty-five generations. We never quite trusted the earlier twenty-five generations because the county magistrate from whom the modern genealogy descends had moved a thousand miles to take his post. Even so, like most Chinese children, I grew up with a sense that my culture and my family had been around for a very long time.

Buddhism, a religion that began in India during the sixth century B.C., had spread to China by the year A.D. 100. Multistory pagodas marked the location of Buddhist temples and monasteries. Buddhism became one of China's three major faiths (along with Confucianism and Taoism), and the Chinese saw no conflict in following the practices of all three.

Confucius

The most influential person in Chinese history was born around 2,500 years ago. His family name was Kong, and his disciples gave him the title Fuzi, which means "master." Kong Fuzi is better known by the Latinized version of his name—Confucius.

In Confucius's time China was divided into warring states. Confucius looked back with fondness on an earlier era when a wise ruler had brought peace to China. He developed a philosophy that stressed the values of duty, harmony, and respect for authority. If each person in society followed the duties appropriate to his or her position, he thought, then society would operate in a harmonious fashion. Peace and prosperity would follow.

Confucius described five basic relationships within society. These were: father/son, husband/wife, older brother/younger brother, friend/friend, and subject/ruler. Each of these except friend to friend was a relationship of superior to inferior. Sons, for example, owed respect and obedience to their father. The relationships were not, however, one-sided. The father also had to care for his children and wife. The subjects of a ruler must follow his commands, but he in turn must govern wisely.

Confucius spent his lifetime teaching his ideas to others. He wanted to train his students to become *junzi* (gentlemen), so that they could change society. However, Confucius was never able to realize his dream of finding a ruler who would accept his ideas and use them for governing. He died in 479 B.C., believing that his life's work was in vain.

Four centuries later, an emperor proclaimed Confucianism as the state religion of China. Its emphasis on authority was used to strengthen the emperor's power. From that time until 1911, the officials of the Chinese government were selected for their knowledge of Confucian thought. They had to pass grueling tests on such subjects as poetry, history, and calligraphy—the hallmarks of a gentleman.

Over the centuries Confucian values permeated the daily life of the people of the world's largest nation. Today, visitors still come to pay homage at Confucius's home and grave. His ideas remain a strong force among Chinese at home and among their overseas descendants.

American handbills flooded China beginning in 1848. This one was circulated by a labor contractor.

"There are laborers wanted in the land of Oregon in the United States of America. They will supply good houses and plenty of food. They will pay $24 a month and treat you considerately when you arrive. There is no fear of slavery. All is nice. The money required [for the voyage] is $58. Persons having security can have it sold, or borrow money of me upon security. —Ah Chan"

THE DECISION TO LEAVE

From the 1850s to the present day, tales of the wealth to be made in America inspired Chinese to emigrate. In one of the earliest Chinese accounts of coming to America, Huie Kin describes his decision, at age 14, to leave his homeland in 1865.

In school we did not study geography. We had only the venerable classics to commit to memory. So the outside world was just a vague notion in my mind, and, for that reason, all the more fascinating. Once a cousin...returned from Chinshan, the "Gold Mountain," and told us strange tales of men becoming tremendously rich overnight by finding gold in river beds. To this day, San Francisco is known among our people as the "Old Gold Mountain." Once I was very sick, and

Deciding to emigrate often caused a young man to contemplate his village and the hills where he had roamed as a boy. If he set out for the Golden Mountains of America, he might return wealthy and prosperous. Or he might never again see his family and the only place he had ever known as home.

in my delirium, so mother told me, I talked of nothing but wanting to go to Chinshan.

Huie Ngou, Huie Yao, Huie Lin and I were good chums as well as cousins. We studied in the same village school, played shuttlecock on the village green, and spent long days together in the foothills with our cows. We told each other our woes and joys, and concocted ambitious exploits, one of which was to go together across the great sea to that magic land where gold was to be had free from river beds and men became rich overnight....

We knew what poverty meant. To toil and sweat year in and year out, as our parents did, and to get nowhere; to be sick and burn or shiver with chills without a doctor's care; always to wear rough homespun; going without shoes, even in cold winter days; without books or time to learn to read them—that was the common tale of rural life, as I knew it.

In 1906, Lee Chew, who owned a laundry in New York City, looked back on his 20 years in the United States. He remembered the first time he thought of going to America. His imagination was stirred by the triumphant homecoming of a man from his village who had actually gone there.

This man had gone away from our village a poor boy. Now he returned with unlimited wealth, which he had obtained in the country of the American wizards. After many amazing adventures he had become a merchant in a city called Mott Street, so it was said. [Mott Street was the principal street of New York City's Chinatown.]

[On returning home, the man] built a palace and summer house and about twenty other structures, with beautiful bridges over the streams and walks and roads. Trees and flowers, singing birds, water fowl and curious animals were within the walls.

When his palace and grounds were completed, he gave a dinner to all the people.... One hundred pigs roasted whole were served on the tables, with chickens, ducks, geese and such an abundance of dainties that our villagers even now lick their fingers when they think of it....

The wealth of this man filled my mind with the idea that I, too, would like to go to the country of the wizards and gain some of their wealth, and after a long time my father consented, and gave me his blessing, and my mother took leave of me with tears, while my grandfather laid his hand upon my head and told me to remember and live up to the admonitions of the Sages, to avoid gambling, bad women, and men of evil minds, and so to govern my conduct that when I died my ancestors might rejoice to welcome me as a guest on high.

Wild geese fly over the sea
My love for China will always be
But war and famine made
her grieve
When the house is poor
The sons must leave.

—"Eight Pound Iron," Charlie Chin

Civil wars, high taxes, natural disasters, and foreign invasions all afflicted China during the 19th and 20th centuries. The result was poverty, famine, and the sound of children weeping for food that their parents could not provide.

"Americans are very rich people. They want the Chinaman to come and will make him welcome ...There will be big pay, large houses, and food and clothing of the finest description. You can write your friends or send them money at any time, and we will be responsible for the safe delivery. It is a nice country, without mandarins or soldiers. All alike; big man no larger than little man. There are a great many Chinamen there now, and it will not be a strange country. China God is there, and the agents of this house. Never fear, and you will be lucky. Come to Hong Kong, or to the sign of this house in Canton and we will instruct you. Money is in great plenty and to spare in America. Such as wish to have wages and labor guaranteed can obtain the security by application at this office."

—A copy of a circular issued in the Chinese language and sent into the country around Canton by a Chinese broker's establishment in Hong Kong, who represented the foreign shipmasters

In 1966, Mao Zedong, the leader of the People's Republic of China, launched a Cultural Revolution to restore the revolutionary spirit of the Chinese Communist party. Zealous young people, known as Red Guards, paraded through the streets carrying revolutionary slogans.

Wong Chun Yau was one of those who fled the communist regime established in China in 1949. Her decision to leave came about because of persecution by the Red Guards, fervent young communists who were part of Mao Zedong's Cultural Revolution.

In China, I owned two houses. And if you had money in China, it was a crime. If you were an intellectual, it was a crime. The really poor, who didn't have a thing, they were the average, so no harm came to them. But if you had a cent, they would purge you. If you owned land, they would purge you. I was purged by the Red Guards twice. This was in the 1960s. Every time they had some movement, they would drag me out, and make me the center of the event. They took everything—my money, my furniture.... They beat me—took off my jacket and beat me. I was sick for three months after that. They stuck me in a cow pen. And then they kept telling me to list my crimes. In the mornings, when I got up, I would have to write. But what could I say? I didn't kill anyone, or set any fires. So what was I supposed to write? So they told me to write down all the things I did against humanity. But I couldn't figure out what I did against the people. I was never a thief or anything. It was a very painful period.

I worked as a nurse for over twenty years. From seven to five I would go to work. But then from seven to nine in the evening, I would have to go to class to learn about the Party and communism. I was envied by a lot of people where I worked, because I was making over eighty dollars a month. And those who were new were making maybe thirty dollars.

Some of today's Chinese immigrants come from Vietnam and other countries of Southeast Asia. For centuries, Chinese migrated to neighboring countries in search of greater economic opportunity. When North Vietnam conquered South Vietnam in 1975, almost 2 million ethnic Chinese lived there. The victorious Vietnamese treated the minority Chinese population cruelly, and many fled. A young refugee who graduated from a Seattle high school tells his story.

I am a foreigner who has grown up in a mass of fire and war called Vietnam. My family was not rich but we were happy. In 1975 the Communists took away South Vietnam and with it went my future hopes and happiness. I could not endure living under Communist control so I was determined to escape in search of freedom.

"When officials, whether soldiers or civil servants, illegally go out to sea, to trade or to settle on islands there to live and farm, they shall be considered as conniving with rebels, and if caught shall receive the death penalty. Magistrates found conniving in such an offense on the part of others likewise shall receive the death penalty.... Any official responsible for the arrest of ten illegal emigrants shall be accorded one merit toward his promotion; if one hundred such culprits, his reward shall be promotion to the next higher rank."

—*From* The Laws and Precedents of the Ching Dynasty, *volume 20*

The missing person in this photograph is the father of the family. He probably received the picture in America as a reminder that his family waited eagerly for his return.

In 1862 and 1863, Liu Chang-yu was the governor-general of the provinces of Kwangtung and Kwangsi. He played a role in the harsh suppression of the Taiping Rebellion, in which 20 million people died. The devastation caused by the fighting caused many Chinese to immigrate to the United States.

Chinese immigrants often brought their own cooking utensils and food on the ships to America. The food that the shipping companies provided for them was usually of poor quality, and the immigrants preferred the rice, noodles, and vegetables that they were accustomed to eating.

CHAPTER TWO

VOYAGE TO AMERICA

Setting out on the 7,000-mile journey across the Pacific was a terrifying experience for young men and boys who had never been more than a few miles from their homes. The *gum shan haak,* or "traveler to the Golden Mountains," faced unknown dangers and hardships. Farewells were often sorrowful, for no one knew if the emigrants could keep their promise to return someday. A folk song from south China expresses the feeling of a wife left behind:

Flowers shall be my headdress once again,
For my dear husband will soon return from a distant shore.
Ten long years did I wait
Trying hard to remember his face
As I toiled at my spinning wheel each lonely night.

Yet the dream of a better life was strong enough to attract many young men. Most brought nothing but the clothes on their backs and a willingness to work. Immigration records from the 1800s are incomplete, but they show that more than 60,000 Chinese came to the United States between 1850 and 1860. The *gum shan haak* set out with courage and hope and little real idea of what hazards they would face.

From their native villages, the emigrants traveled in flimsy junks or rafts down the waterways of the Pearl River to Hong Kong. In the great harbor there, the farm boys were awed by the sight of ships larger than any houses in their village. Many saw white people there for the first time in their lives. As Huie Kin recalled, "to my rustic eyes, they certainly appeared a strange people, with fiery hair and blue-gray eyes. It was not to be wondered at that we nicknamed them *hung-mao-kuei* ("red-haired fellows"), an appellation that was meant to be merely descriptive and not offensive."

The sojourners' first difficulty was paying for their passage. In the 19th century, a ticket from Hong Kong to San Francisco cost between $25 and $60, depending on the comforts provided on board ship. This was often more than a Chinese family's annual income. To pay for the sea voyage, some saved their hard-earned money or sold livestock and property. Others borrowed from relatives or money-lending associations called *hui.* The debt was a family responsibility; if the emigrant did not return with his hoped-for wealth, his parents, wife, and children had to work to repay it. All their hopes were riding with the son or husband who promised that some day he would return.

Other ways of paying for the overseas passage soon developed after the first wave of Chinese immigrants proved that they were hardworking, faithful employees. Charles Crocker, the head of the Central Pacific Railroad, found that his best workers were Chinese. He sent agents to Chinese villages to recruit more laborers. Other companies soon followed, flooding China with handbills that promised "good houses and plenty of food" for workers who signed up. The "contract laborers" signed contracts that required them to work for a certain time at a fixed rate of pay, in return for travel expenses.

Chinese merchants caught on to the system and began to sell "credit tickets." The emigrants agreed to pay back the money, with interest, after they found jobs overseas. Assisting emigrants was so profitable that it gave rise to businesses called *gum shan chong,* or "Golden Mountain Firm." These were import-export firms that carried on trade between China and the United States. They often helped to process paperwork and book passage for emigrants. Some provided facilities where emigrants could stay in Hong Kong or Canton

The Chinese endured seasickness, poor food, crowded conditions, and loneliness.... Days and days of seeing nothing but the endless ocean created the fear that they would never reach shore again.

while waiting for their ship to depart. Many Chinese were victimized by the terrible "coolie" system. The word *coolie* may be derived from the Chinese *kuli,* which means "bitter strength." A coolie was virtually a slave who was required to work at jobs no one else would take, such as laboring on the sugar plantations in Cuba. Most coolies were tricked into signing labor contracts, and some were simply kidnapped and taken aboard ships against their will.

It is important to note that virtually *no* Chinese who immigrated to the United States was a coolie—though the word was often used as a term of contempt for all Chinese immigrants. In fact, most came of their own accord, although later in the 19th century, some young Chinese women were forcibly brought to this country as prostitutes.

In 1853, an agent of a British company in China described the departure of a ship from Hong Kong carrying "1,400 Chinese laborers...to work for an American railway company.... The vessel had got under way amid the firing of crackers and the uproar of gongs and drums as a token of the emigrants' satisfaction." The British agent may not have known that Chinese set off firecrackers to scare away evil demons.

Even so, the voyages were long and miserable. The Chinese endured seasickness, poor food, crowded conditions, and loneliness on their way to a new life. Even those who had been fishermen had seldom ventured far from land. Days and days of seeing nothing but the endless ocean created the fear that they would never reach shore again. Almost none could speak any language but their native dialect, and it was impossible to communicate with the crew or officers. Shipowners regarded their steerage passengers merely as cargo and made few concessions to their comfort.

Though the small Chinese fishing boats, known as junks, were not suited for ocean voyages, a few of them actually crossed the Pacific. Two reached the California coast in 1854. The one shown here arrived in 1924.

Today, immigrants from Taiwan and Hong Kong have an easier time than their 19th-century forebears. Jet planes cross the Pacific in a few hours, bringing newcomers to gleaming airport terminals. Often, good jobs await them, for many are college graduates skilled in modern science and technology.

Even so, other modern Chinese immigrants faced hardships as great as any of the farmers' sons who first came here more than a century ago. These are the boat people of Southeast Asia, who fled their homelands because of anti-Chinese discrimination. Their journeys often began in a flimsy fishing boat never intended for travel on the open sea. Overloaded with passengers, the boats sometimes sank before reaching port. Pirates preyed on them during the voyage, taking the few possessions the helpless refugees had. No one knows how many boat people died in the attempt to reach freedom.

The hope of a better life still compels people to set out for the United States. The dream of Gum Shan, "the Golden Mountain," is still alive.

A group of Chinese prepare to leave their ship at Honolulu. After 1850, merchants from the United States brought Chinese to work on the islands' sugar plantations.

Where the sun rises
In the land of Fu Sang
There is my home
Seeking fame and riches
I came to the land
Of the eternal flowers.

—Anonymous

GETTING READY

Once Huie Kin decided to go to Gum Shan with his cousins, he faced the problem of making his dream a reality. As he recalled:

It cost a lot of money to go abroad; thirty dollars was the minimum steerage fare. I had seen these white, round discs of silver ["trade dollars" minted in the United States for use on overseas trading voyages]. When father was unusually successful in selling the farm crop in the market town, he would come home with a few pieces and exhibit them to us, knocking them together to produce a pleasing, musical sound. That was the way people told good coins from the bad, for the bad ones would not give the mellow tone when struck. They would either be mute or have a sharp metallic sound. On one side of the coin was a woman standing with some flowers in one hand, and so we called them the *jar-far-ngan,* "flower-holding silver." I believe that they were imported from America

A copy of a labor contract, written in both Chinese and English, that obliged workers to pay a part of their wages each month to cover the cost of their ship ticket. The date was the 29th year of the current emperor's reign—or 1850.

英商華民合同立議約、今有祥勝行、特僱花旗國船、名啞嗎三、裝運自欲出
洋傭工之人、駛往加拉吠呢啞國、咈蘭噺戈口、代爲尋覓生理、自
上海起行、一應伙食船鈔等費、俱係祥勝行東家、代爲應付、到彼
處、尚需東家代薦生理、其代付之銀、理應歸還、俟生理定確、即向
本東家、預支伙食船鈔水脚洋銀、每人壹伯貳拾五元、交還祥勝
行東家親收、即向該處僱爲作工之商議定、每月扣去工金若干、
待一切扣清、方照月付銀、此係兩願、不得異言、今欲有憑、立此議
約、各執一紙爲照

英商
華民

己酉年 月 日立

AGREEMENT
BETWEEN
THE ENGLISH MERCHANT & CHINAMEN.

The *Tseang Sing* Hong having now hired the American Ship called the *Ah-mah-san* for voyaging purposes, the mechanics and labourers, of their own free will, will put to sea, the ship to proceed to *Ka-la-fo-ne-a*, and port of *Fuh-lan-sze-ko*, in search of employment for the said mechanics and labourers. From the time of leaving Shanghae, the expenses of provisions and vessel are all to be defrayed by the head of the *Tseang Sing* Hong. On arrival, it is expected that the foreign merchant will search out and recommend employment for the said labourers, and the money he advances on their account, shall be returned when the employment becomes settled. The one hundred and twenty-five dollars passage money, as agreed by us, are to be paid to the said head of the said Hong, who will make arrangements with the employers of the coolies, that a moiety of their wages shall be deducted monthly until the debt is absorbed: after which, they will receive their wages in full every month.

The above is what we agree to, and there must be no differing words; and as evidence, we enter into this contract, a copy of which, each party is to have.

Done in the Moon of the 29th Year of Taou Kwang.

by the early traders as a medium of exchange.

One day our plan was completed and we were to tell our fathers that we had decided to go to America and wanted them to buy the tickets for us. I worried over it for several nights, not knowing how to approach father without appearing ridiculous in his eyes, for I was only fourteen and he was poor. When finally I did muster enough courage to approach him, to my great surprise he raised no objection whatever; and, as he had no ready cash, he borrowed the thirty dollars from a well-to-do neighbor, with our little farm as security. Father said very little about the matter during the intervening days before my departure; but I think that he had quite a good business head, and looked upon the thirty dollars as a good investment, for he, too, had talked with returned cousins from abroad, and every one of them went away poor as himself but came back with the beautiful gold pieces in their pockets. Probably he had also dreamed of going abroad, but he was married and had a family on his hands. His son was plucky to want to go, and he might be equally lucky as the other cousins; then they would not have to toil and struggle any more.

This man's face shows the strength and determination of the Chinese sojourners who set out for Gum Shan. By Chinese law, all males had to wear their hair in a queue, or pigtail. Without it, they could not return home.

In 1958, Chung Kun-ai remembered the preparations for the trip he made from China to Hawaii—79 years before. He was one of many "contract laborers" who made the journey to what the Chinese called the Sandalwood Islands.

How did these poor villagers from South China raise the money to make these long and perilous journeys, you may well ask? They all had the most difficult time to make both ends meet out of tilling the soil in their native villages.... Of course, [in Hawaii] we know that the plantation companies supplied money to their Chinese recruiting agents who went into the villages of the Kwangtung seacoast to sign up contract laborers for the plantations. The first of these contract laborers arrived in Honolulu in 1852. Other groups followed. All these migrants had their passage paid for by the plantations, so that financing a sea journey was no problem for them. In the colorful colloquial language of the day, those who signed on as contract laborers were labelled as villagers who had sold themselves as "pigs." The smaller group that came on their own either had parents who had the money, or borrowed money from others at home or abroad, to be repaid out of their earnings. At least, that was what these migrants promised to do when they borrowed money—to return the money at interest.

SAYING GOOD-BYE

Writing poetry, one of the skills Confucius stressed for "gentlemen," allowed Chinese immigrants to express their feelings of loneliness. One sojourner, writing in San Francisco in 1911, remembered his departure from home:

I bid farewell to Father and Mother.
Throughout the journey I worry and worry about them.
To get food, I have no choice but to hurry about;
My thoughts of them are all in a tangle, like hemp fibers.
Unable to return—
When will I again sit beside their knees?
An endless horizon severs family happiness.
As I long in vain for my parents, my eyes turn blurry.

The day of departure for America was often filled with painful farewells to relatives and friends. Huie Kin, leaving in 1868, was too excited to worry.

Huie Ngou, Huie Yao, Huie Lin [his cousins], and I got ready for the journey. Ngou was twenty-five and, being the oldest, was made leader of the little party. Yao and I were fourteen and Lin was twenty. One Spring morning we started out before daybreak so that people might not see us go. In this way we avoided any possibility of hearing unlucky words spoken at our departure, thus enhancing the probability of our having a safe voyage. Our baggage consisted of a roll of bedding and a bamboo basket with netting on the top. Into this we put our shoes, hat, and all our worldly possessions, including some homemade biscuits that mother had put in for the journey. Curiously enough, I was not a bit sorry to leave home or my people. We said good-by at the doorsteps, and a minute later darkness closed around us and we could not see the folk standing there. Only a few days later, while waiting idly in Hongkong, did I get homesick, and thought of some books I had left in the school desk and the kite hanging on the wall in the bedroom.

Sometimes the parting was particularly sad. Nearly half a century later, Mrs. Teng still recalled how her mother sold her to raise money for her father's funeral expenses.

My mother took off my mourning robes, dressed me in a colored dress with a red string on my hair. I went with this lady [a go-between] to the big house of Mr. Chin, two miles from our village. He was to look me over and I seem to be his choice for he took out ninety dollars to give to my mother. Every year in my age was worth ten dollars. I wished I were older than nine so that my mother could get more money.

Before the actual parting I was happy and glad to go because I knew I was helping mother. When my mother and me went out of the house I took a long look behind and did not want to go. I cried and begged and asked to stay at home. For once I had the sympathy of the neighbors. They cried and told me that I must be a good girl and go so that my mother can beg the money to pay [for] the coffin.... It was eleven o'clock when we came to the gate of Mr. Chin's house. We stayed outside until it was twelve. It is said that it is bad luck to enter a master's house when the time is odd, it must be even time. Again the parting was hard. I ran after my mother but my master held me. He gave me a silver spoon, a jade bowl, sweets, and cakes—all that I always longed for. I was glad to stay forever.... I was so poor for a long time that those sweet and pretty things took a great hold on me.

A lady in that house told me that Hawaii had big, fat, very sweet sugar cane—it was better than honey. I was so crazy for cane that I just waited for the day to come to Hawaii. She also told me that there was hardly any [work] to do but after I came I found that this was not true.

In 1972, Lilac Chen, an 84-year-old woman, still remembered bitterly that her own father had sold her when she was only six.

My worthless father gambled every cent away, and so, left us poor. I think my mother's family was well-to-do, because our grandmother used to dress in silk and satin and always brought us lots of things. And the day my father took me, he fibbed and said he was taking me to see my grandmother, that I was very fond of, you know, and I got on the ferry boat with him, and Mother was crying, and I couldn't understand why she should cry if I go to see Grandma. She gave me a new toothbrush and a new washrag in a blue bag when I left her. When I saw her cry I said, "Don't cry, Mother, I'm just going to see Grandma and be right back." And that worthless father, my own father, imagine...sold me on the ferry boat. Locked me in the cabin while he was negotiating my sale. And I kicked and screamed and screamed and they wouldn't open the door till after some time, you see. I suppose he had made his bargain and had left the steamer. Then they opened the door and let me out and I went up and down, up and down, here and there, couldn't find him.... Then a woman from San Francisco...picked me up and brought me over.

When a young man left his village to seek a better livelihood in a faraway land, he often left a young wife behind, with the promise that he would return someday. The parting was not a joyful one. A Hakka folk song of that period expresses a bride's sorrow as she calls across the hills after her husband:

I beg of you, after you depart, to come back soon,
Our separation will be only a flash of time;
I only wish that you would have good fortune,
In three years you would be home again.
Also, I beg of you that your heart won't change,
That you keep your heart and mind on
taking care of your family;
Each month or half a month send a letter home,
In two or three years my wish is to welcome you
home.

This photograph was found in the Bayard Street apartment of an old man in New York's Chinatown. It was probably his only memento of the village in China where he was born. The high proportion of women and children indicates that many men had left to seek their fortunes elsewhere.

GOING TO AMERICA

Huie Kin, like many of the early immigrants, went down the Pearl River to the British colony of Hong Kong to board a ship for America.

Finally, the day was set for the ship to sail. People called it the Wing Tung Ki ship, but I do not know whether it was the name of the boat or the name of the agents that chartered her for the trip. It was a big sailing vessel, with three heavy masts and beautiful white sails, which looked especially well when the wind was behind and the sea was smooth. The captain and the sailors were all white people. Quite a number of Chinese passengers went on board. We were lined up on the deck for the inspection of our tickets. The officer counted noses and tickets to see that the numbers agreed. A small boy was hidden in one of the bamboo baskets with netting on the top, and so escaped payment of his fare. He was the only stowaway on our boat.

We were two full months or more on our way. I do not know what route we took; but it was warm all the time, and we stopped at no intermediate port. When the wind was good and strong, we made much headway. But for days there would be no wind, the sails and ropes would hang lifeless from the masts, and the ship would drift idly on the smooth sea, while the sailors amused themselves by fishing. Occasionally, head winds became so strong as to force us back. Once we thought

Easily preserved when dry and simple to cook, rice remained among the favorite foods of Chinese wherever they went. The use of chopsticks—called kwaizi, *or "quick ones"—was another Chinese custom.*

An artist for an American magazine sketched Chinese preparing to depart for the United States around 1850. The drawing shows little trace of the stereotypes that were typical of later depictions of Chinese immigrants.

Chinese on board a sailing ship bound for Honolulu in 1900. Immigrants to Hawaii endured less racism than those on the U.S. mainland. By the turn of the century there were more than 150,000 Chinese in Hawaii, about one-sixth of the population.

we were surely lost, for it was whispered around that the officers had lost their bearings. There was plenty of foodstuff on board, but fresh water was scarce and was carefully rationed. Not a drop was allowed to be wasted for washing our faces; and so, when rain came, we eagerly caught the rain water and did our washing.

One morning Huie Ngou [the leader of the party] was suddenly taken sick with fever, and we woke up to find him gasping and choking for breath. He passed out that afternoon, and his body was wrapped in a sheet and quietly lowered into its watery grave at night. For hours afterwards, we stood by the side of the ship, gazing into the darkness, scarcely knowing what to think. Were I older, I might have indulged in philosophical reverie as to the brevity and vanity of human existence.... But the thought uppermost in our youthful minds was a practical one—What was to become of the party now that our leader was gone?—and there was an uneasy feeling that his death could not but cast an evil shadow upon our venture. Years later we heard that Ngou's ghost went back to his old village and haunted a woman who claimed that we had thrown him overboard when he was still alive.

Conditions aboard the 19th-century immigrant ships were crowded. The passengers slept on narrow bunks below decks and had to collect rainwater to wash themselves. Disease sometimes spread through the holds where immigrants endured the two-month voyage.

Chung Kun-ai, headed for Hawaii with his father, had a shorter and more pleasant trip.

After two or three days, I got over my seasickness and began to enjoy the voyage. The ocean in time calmed down also, and it remained smooth sailing for the rest of the voyage of forty or more days. I remember that we were in all about three hundred passengers on board. I was a growing lad, and the salty sea air whetted my appetite. Father believed in taking just two meals a day, but at noontime some of the other passengers took their noonday bite of bread boiled in water. My mouth watered. Fortunately those who were having lunch were kind enough to share their bread with me.

Well, a sea voyage in those days dragged on and on. Once in a great while, some excitement changed the monotonous routine. For instance, we sailed one day into a school of skipjacks. The sailors tied pieces of white cloth to their hooks as bait and caught a few. Ordinarily we would not consider the skipjack a good eating fish, but on that occasion it proved a delightful and delicious variation in our diet. We enjoyed the skipjacks.

Lee Chew recalled his voyage from China to San Francisco sometime around 1882.

My father gave me $100, and I went to Hong Kong with five other boys from our [village] and we got steerage passage on a steamer, paying $50 each. Everything was new to me. All my life I had been used to sleeping on a board bed with a wooden pillow, and I found the steamer's bunk very uncomfortable, because it was so soft. The food was different from that which I had been used to, and I did not

Today the desire to come to the United States is so great that many enter the country illegally.

An official of the Immigration and Naturalization Service describes the journey of an illegal immigrant from the People's Republic of China:

"Typically, what happens is this: you are a worker in Fujian Province, you go to the head of the Communist party in your village, and he fixes you up with a smuggler. That person gets you a Chinese passport and an exit permit. At Hong Kong, you go to a safe house and someone else will get you the right kind of documentation for the rest of the trip. You can pass through up to ten arrangers, each responsible for one part of your journey. Usually, after Hong Kong, you go to Bangkok....

From Bangkok, it depends which smuggler you hook up with what route you take. It's not your free choice. It depends on how much you pay, and what level of service you're buying. If you avoid Latin America, your trip is much less difficult, and faster. So those with money hit Canada first, and apply for refugee status. Canada is very generous, just like us.

If you have little money, you take the long route. You may go from Canton to Hong Kong to Moscow to Santa Cruz to Managua and by bus or foot across the Mexican border. That is fairly quick—sixty days. There are even longer routes."

This overloaded boat carrying refugees from Vietnam in 1978 sank before it reached shore, but most of the passengers were rescued.

like it at all. I was afraid of the stews, for the thought of what they might be made of by the wicked wizards of the ship made me ill. Of the great power of these people I saw many signs. The engines that moved the ship were wonderful monsters, strong enough to lift mountains. When I got to San Francisco...I was half starved, because I was afraid to eat the provisions of the barbarians, but a few days' living in the Chinese quarter made me happy again. A man got me work as a house servant in an American family, and my start was the same as that of almost all the Chinese in this country.

Mrs. Teng had the longest voyage of all. But at the end, she was disappointed.

In 1891 my master and me sailed on the *Billy Jack* to go to my new mistress in Hawaii. We slept on canvas cots and had cheap meat and cabbage for every meal. We could not land in Honolulu because there was a small pox on board ship. We went directly to San Francisco and stayed there for two months. I never saw the shape of the land for I was below the [deck of] the ship. When we came back to Hawaii I was locked in the immigration office for three weeks. How happy I was when my boss came to me.... The first thing I asked my master was a piece of sugar cane. He said that there is none around the place where we live. How sad I was for I expected cane to be all around.

The voyage to America in the 20th century was sometimes desperate. Quan Ngo, a Chinese who grew up in Vietnam, found life under the new communist regime unbearable. He was among the "boat people" who risked their lives by taking rickety boats into the rough waters of the South China Sea. Quan Ngo wrote the following account for his English class in a Seattle high school.

I remember the night we left. A dull moon filled the lake and little stars twinkled brilliantly in the sky. The night wind swept the leaves to and fro. It seemed like nature was playing a farewell song for us. The fishing boat which we were to use for our escape lay waiting on the water. We left quickly and while we were leaving I waved goodbye to my adopted homeland, Vietnam. "Goodbye my relatives and goodbye my [girlfriend Ming]." Before I went away I didn't want to say a word to her because I was afraid the tears would never leave her eyes. I could not bear seeing those tears on her face. The ocean was freezing and pitch black. I could hear its mighty sounds coming from the bottom of the ship. There were more than 100 people on the boat though it was only twenty-two meters long and six meters wide. I had to sit by the side of the boat on top of an oil drum. If I wasn't careful, I would have fallen overboard. Early the next morning the sun was burning brightly. Each movement of the boat took me further and further away from all that I had known and loved. The whitecaps in the water reminded me of farewell tears. Ming...forget me! Hate me! I am sorry! The boat traveled a day before we reached international waters. During the night I woke up with a strange feeling. The boat was shaking. It started to rain and continued throughout the day. The waves were getting higher and stronger. The ocean turned from blue to gray. The waves attacked our vessel like wild dogs. The ship was silent and everybody was praying. The water came over the rail and hit me in the face. It felt like someone was slapping me. Suddenly, the owner of the ship came out of the control room. He wore a very pale expression. His body was shaking. He had terrible news! The captain was sick and he needed someone who knew how to pilot the ship. The rain was getting harder. It seemed like the sun was gone since the ship looked so gray. We felt very close to death and we prayed to God to help us. The captain recovered and everyone took a long breath. All of a sudden we saw a black shape coming toward us. "At last!" we thought, "Someone is coming to rescue us." Everyone began to cheer up. As the black shape became bigger and closer, we also heard gunfire. Someone was shooting at us! People were afraid. The captain shouted, "We must escape. Throw everything overboard!" Everyone threw their things into the water. The little packages flowed alongside the ship until the waves claimed them and swept everything away. The black ship came closer and closer, still shooting at us. Toward the end of one's life, some people curl up their body, others just cover themselves...I couldn't even think toward the end of my life. I sat still on the oil drum. I just wanted to laugh...laugh at this sour and bitter life. We waited for the robbers to come but the storm was too great even for them. We were saved. The sky turned from gray to black. It was extremely cold. During the attack the drinking water was thrown overboard. The angry sound of the waves mixed with the crying of a little child. Two days later we were still without water and everyone began to lose hope. It was a slow and painful torture. All aboard the boat were waiting to die. With dawn came a beautiful morning mist. I beckoned the ocean to take a message to my relatives and to Ming: Freedom, no matter how much we have to sacrifice, is worthwhile! Suddenly a dark shape came toward us again. Should we go ahead and signal for help? The captain sighed, "Let us gamble for our life. We have nothing to lose." Everyone felt so weak but their restless eyes and dry lips reflected the bitterness of having their lives decided by chance, as if in a game. Perhaps that is what life is: a series of gambles. The dark ship became bigger. There was no gunfire and everyone felt relieved. Gradually, as the ship came closer we could see its colors clearly. It was a big military ship from Malaysia. Everyone rejoiced as it stopped next to us. Even the elderly smiled like little children. I felt I was smiling too. I remember talking but now I can't recall who I was talking to or what we were saying. Life is very precious. Once you are close to death and still are able to live, life has a greater value. Even though we have much pain and hardship, life is worthwhile if we are free.

Yung Wing

Born in 1828, Yung Wing was educated in a missionary school, where he showed great promise. When he was 12, he left China to attend a high school in Connecticut. His ship took the westerly route, through the Indian Ocean. Yung Wing remembered that a terrible storm blew up as the ship was rounding the Cape of Good Hope at the southern tip of Africa. "The tops of the masts and ends of the yards were tipped with balls of electricity. The strong wind was howling and whistling behind us like a host of invisible Furies. The night was pitch dark and the electric balls dancing on the tips of the yards and tops of the masts back and forth and from side to side like so many infernal lanterns in the black night, presented a spectacle never to be forgotten by me."

Yung Wing survived the experience to become, in 1854, the first Chinese graduate of Yale University. He returned to China, where he became a tea merchant. In 1870, he established the Chinese Educational Mission, which for many years sent Chinese students to the United States to receive an education.

The classroom in the Connecticut school for Chinese children that Yung Wing founded. Chinese teachers taught the students their own language as well as English.

Even in 1924, when this ship arrived at the Angel Island Immigration Station, many Chinese were still attempting to enter the United States. The crowd's Western-style clothing indicates that some were merchants who were returning after a visit to their families in China.

CHAPTER THREE

ARRIVAL IN THE LAND OF THE FLOWERY FLAG

No one can say when the first Chinese arrived in what is today the United States. Almost certainly, it was before anyone had ever dreamed of a country with that name. During the 17th century, Spaniards brought skilled Chinese workers to the New World from the Spanish colony in the Philippines. Many of these Chinese converted to Christianity and took Spanish names. One of them was among the 23 founders of the city of Los Angeles in 1781. Though he had the very un-Chinese name of Antonio Rodriguez, the Spanish records list him as a "Chino," Spanish for Chinese.

In 1781, after the United States won its war of independence, the very first trading voyage from the new country went to China. American ship captains were soon using Chinese as crewmen. Three of these sailors—Ashing, Achun, and Accun—were stranded in Baltimore when the captain of their ship left to get married. Though they returned home about a year later, they were probably the first Chinese residents of the United States.

In the early 19th century, American Christian missionaries went to China. Beginning in 1818, they sent some young Chinese boys to a school in Connecticut. Among them was Yung Wing.

However, the vast majority of the early Chinese immigrants arrived in San Francisco. The first three arrived in 1848, the year before the gold rush began. A local newspaper noted that the "two Chinamen and a Chinawoman ...were looked upon as curiosities by some of the growing town of San Francisco, who had never seen people of that nationality before."

Within two years, Chinese were not so unfamiliar. Hundreds of them had joined the horde of gold-seekers arriving in California. At that time, there were no immigration restrictions; people simply swarmed off the ships and headed for the gold fields. As word of the "Golden Mountain" spread through Kwangtung Province, thousands more joined the gold rush. About 18,000 Chinese arrived in 1852 alone.

Called "Celestials" because China was then known as the Celestial Empire, the Chinese were still unusual enough to cause comment. The men all wore queues, long pigtails that the Manchu government required for every male Chinese. A California newspaper reported: "A very large party of Celestials attracted considerable attention yesterday evening...on their way to the southern mines. They numbered about fifty, each one carrying a pole, to which was attached large rolls of matting, mining tools and provisions.... They appeared to be in excellent spirits and in great hopes of success, judging from their appearance."

Aside from their appearance, there was one important difference between the Chinese and the Germans, Irish, Spanish, or English who also went to California to find gold. The Chinese could not legally become citizens of the United States. A federal law passed in 1790 said that only people of the white race could become naturalized citizens.

This did not seem significant at the time, for the Chinese gold-seekers had no intention of staying. Most hoped only to make money and return home to their families. However, that hope proved a vain one for many Chinese. Like thousands of other hopeful arrivals, they went bust in the gold rush. The disgrace of returning home without the hoped-for riches kept many of them in the United States, still seeking their fortunes.

Unlike other immigrants, none of the Chinese living in the United States could appeal to a representative of their homeland for help. China's government had no con-

In 1882, Congress passed the Chinese Exclusion Act....
For the first time, the United States closed its borders to a group of people solely on the basis of their race.

suls in the major cities of the United States, as European countries did. Indeed, the Chinese Empire regarded its overseas citizens as "wild geese" who were unworthy of attention.

The Chinese immigrants received some legal protection when the United States and China signed a treaty in 1868. Called the Burlingame Treaty after the American ambassador to China, the agreement allowed American and Chinese citizens to migrate freely "from one country to the other." Chinese were permitted to become "permanent residents" of the United States—but still not citizens.

The Burlingame Treaty brought a second wave of Chinese immigrants, larger than that caused by the gold rush. Those Chinese who had returned home—usually, the successful ones—told stories of the great opportunities available in the "Land of the Flowery Flag." (The stars on the U.S. flag looked like flowers to the Chinese.) Between 1860 and 1880, the Chinese population of the United States tripled, to about 100,000 people.

Still, the Chinese faced vicious prejudice from people who feared they were taking jobs away from "white Americans." Political pressure caused the U.S. government to renegotiate the terms of the Burlingame Treaty. In 1882, Congress passed the Chinese Exclusion Act, surely one of the most shameful laws in American history. It prohibited the immigration of Chinese laborers for 10 years but was later extended for an indefinite period. For the first time, the United States closed its borders to a group of people solely on the basis of their race.

There were some exceptions to the Chinese Exclusion Act. Students, merchants, tourists, and diplomats were permitted entry. Also, those who had previously established themselves as "permanent residents" could leave the country and return—though they were prohibited from bringing wives or other family members to the United States.

Duplicate

WHEREAS :---------Haw How Ming---------- whose photograph is hereto attached, is about to depart for China intending to return to the United States; NOW THEREFORE, for the better identification of the said Haw How Ming and in order to facilitate his landing upon his said return, we, the undersigned, do hereby certify and declare under oath, each for himself and not one for the other, that we have known the said Haw How Ming as a resident of the City & County of San Francisco, State of California,-------------- for the number of years set opposite our respective names; that he is by occupation a merchant in good standing engaged in buying and selling merchandise at a fixed place of business, to-wit: as a member of the firm of Ho Hop & Company,---------------- dealers in Boots & Shoes,---------------- doing business at # 742 Clay Street,---------------- in said City & County, and that he now resides thereat and does not engage in the performance of any manual labor, except such as is necessary in the conduct of his business as such merchant, and that he has

An 1899 certificate carried by a Chinese American merchant that permitted him to leave the country and return. Two non–Chinese American citizens were required to sign the document, certifying that they knew the man was a merchant, not a laborer.

However, even the "permanent residents" had to provide affidavits, signed by two non–Chinese American citizens, in order to travel abroad. Immigration officials sometimes rejected legitimate documents as forgeries. It became dangerous for Chinese Americans who had lived in the United States for decades to visit their families in China because many were refused admission upon their return.

From this point on, U.S. immigration officials cracked down on Chinese trying to enter the country. All new arrivals faced intense grilling when they passed through the customs station. If they could not prove that they were among the categories exempted from exclusion, they were sent back on the next boat.

However, there was one way for a Chinese to become a legal citizen, entitled to come and go freely. Despite the 1790 law, anyone—of any race—born in the United States was automatically granted citizenship. By 1900, the U.S. census counted 6,657 such Chinese American citizens.

Then, in 1906, an earthquake devastated the city of San Francisco. Chinese residents suffered as much as others from the disaster, but it brought an unexpected boon. Because the fires that followed the earthquake destroyed the office containing birth records, many Chinese "permanent residents" now applied for citizenship, claiming they were native-born. In many cases, their requests were granted.

Moreover, as American citizens, these Chinese were entitled to bring wives and children from China. Suspicious immigration officials noticed a sudden influx of large Chinese families—mostly young men. In fact, many were "paper sons," who had paid a Chinese American "father" to sign

The immigration officials, known as *luk yi*, or "green-clothes men," grilled the immigrants at Angel Island for hours at a time...trying to catch them in a mistake or contradiction.

false birth papers for them. This was their ticket to the United States and the chance for a better life.

Questioning of new Chinese arrivals thus became more severe. Between 1910 and 1940, every Chinese arrival on the West Coast had to go through Angel Island, an immigration station in San Francisco Bay. Unlike Ellis Island in New York Harbor, through which so many European immigrants passed, Angel Island was for Asians only. The newcomers were held there for weeks—and in some cases, as long as two or three years. The immigration officials, known as *luk yi,* or "green-clothes men," grilled the immigrants at Angel Island for hours at a time. The *luk yi* asked countless questions about the immigrants' home villages, the places where they claimed to have been born, trying to catch them in a mistake or contradiction.

Chinese men, as citizens, took advantage of the opportunity to bring their wives to this country. Between 1910 and 1924, the number of female Chinese immigrants rose from almost nothing to more than 1,000 a year. However, even this small number was alarming to those Americans who saw the Chinese as a threat. In 1924, a new federal law barred aliens ineligible for citizenship from entering the country. This included the Chinese-born wives of Chinese American citizens. "We were beginning to re-populate a little now," recalled a Chinese man, "so they passed this law to make us die out altogether."

The anti-Chinese hysteria had exactly that effect. The number of Chinese living in the United States, citizens and noncitizens, steadily declined. It had reached a peak of a little more than 100,000 in 1890; by 1940, there were only about 57,000 people of Chinese descent in the United States. Only two mainland cities, San Francisco and New York, had more than 10,000 Chinese residents.

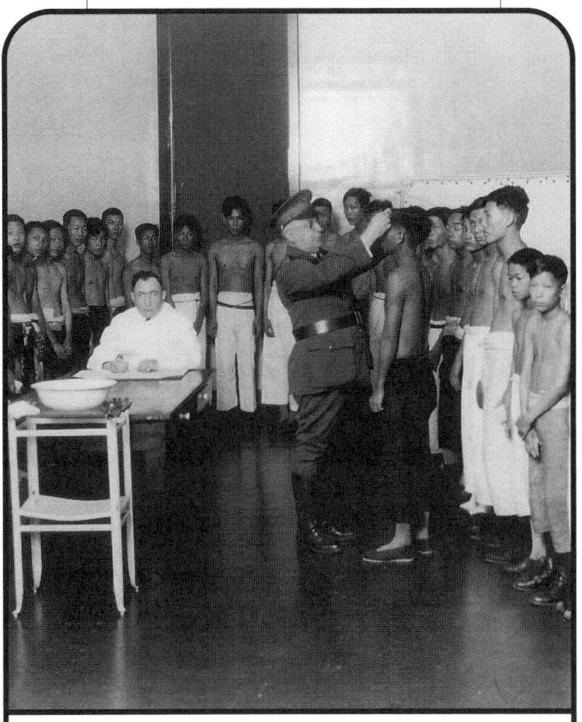

Very likely, many of these young men and boys at Angel Island in 1923 were "paper sons." A physical examination by the luk yi, *"green-clothes men," was among the requirements for entry.*

In 1943, Congress passed a bill permitting Chinese to immigrate—but it set a yearly quota of just 105 people. The law also enabled Chinese who were already living here to apply for citizenship. Though the "golden door" opened only a crack, there was at last no ban on immigrants simply because of race. By now, Angel Island was closed, and Chinese were treated like any other immigrants.

After World War II, the small quota limits were relaxed for two groups of Chinese. Once again, the wives of Chinese American citizens were allowed entry—showing the nation's gratitude to the many Chinese American soldiers who served in the war.

Chinese students made up the second group. After 1945, many of them enrolled in American universities. Then, in 1949, a communist government came to power in China. Some Chinese students in this country chose not to return home, and others were prevented from doing so after the United States entered the Korean War in 1950. Highly educated, they found positions as university professors and in new high-tech industries.

In 1965, a new immigration law opened the "golden door" even wider. It instituted a new quota system that gave preference to people with special talents and skills and allowed up to 20,000 new immigrants from any country. Since then, hundreds of thousands of Chinese have arrived from China, Taiwan, Hong Kong, and the countries of Southeast Asia. The 1990 census showed that nearly 2 million Americans claimed Chinese descent. Though they make up less than 1 percent of the total population, Chinese Americans are no longer "curiosities." They have proudly taken their place in the mainstream of American life.

The little town of San Francisco as it looked in 1847, the year before the first Chinese arrived, and two years before the gold rush changed the city forever.

A FIRST LOOK AT AMERICA

Most early Chinese immigrants arrived at San Francisco. Sixty-four years after landing there, Huie Kin remembered his feelings on the day he first saw America.

On a clear, crisp, September morning in 1868, or the seventh year of our Emperor T'ung Chih, the mists lifted, and we sighted land for the first time since we left the shores of Kwangtung over sixty days before. To be actually at the "Golden Gate" of the land of our dreams! The feeling that welled up in us was indescribable. I wonder whether the ecstasy before the Pearly Gates of the Celestial City above could surpass what we felt at the moment we realized that we had reached our destination. We rolled up our bedding, packed our baskets, straightened our clothes, and waited.

In those days there were no immigration laws or tedious examinations; people came and went freely. Somebody had brought to the pier large wagons for us. Out of the general babble, someone called out in our local dialect, and, like sheep recognizing the voice only, we blindly followed, and soon were piling into one of the waiting wagons. Everything was so strange and so exciting that my memory of the landing is just a big blur. The wagon made its way heavily over the cobblestones, turned some corners, ascended a steep climb, and stopped at a kind of clubhouse, where we spent the night.

In 1849, San Francisco Harbor was choked with abandoned ships. Their crews had deserted to join the search for gold. It was impossible to construct new buildings fast enough to keep up with the surge in population, and tents dotted the streets.

When John Jeong was 84, he recalled his first impressions of the United States.

In 1900 the Immigration Office was not on Angel's Island, it was upstairs on the pier. So when I arrived in this country I had to stay in the place on the pier about two weeks, because they were investigating my case. There was a big room there, for everyone to sleep in, and then a big eating hall with long tables. I remember we ate our meals standing up and we weren't allowed to write letters there. Finally they said I and a few others were all right. They put us in a horse carriage and we drove into Chinatown. It was an open carriage with standing room only. Halfway there some white boys came up and started throwing rocks at us. The driver was a white man, too, but he stopped the carriage and chased them away. I was thirteen at the time.

Steven C. Lo was born in Taiwan, the island nation established in 1949 by the Chinese Nationalists after their defeat in the Chinese civil war. Many recent Chinese immigrants to the United States have come from Taiwan. In his autobiographical novel, The Incorporation of Eric Chung, *Lo describes the arrival of one such immigrant.*

I'd say September of 1972 was when all this started, September 17, 1972, to be exact. My Date of Entry. On that day, as one of a crowd, I got off a Japan Airlines 747 at Los Angeles Airport. I became one of those you'd call "America's new arrivals"—you know, the type of people others like to help a lot. Probably because we were forever grateful for anybody who had time to tell us a thing or two. For the officers at the Immigration, I was a "Non-Resident Alien." From them I received a card that I had to fill out every year on New Year's Day and a sheet that said if I ever looked for a job I'd be "subject to deportation."

It was about five or six in the afternoon when I got off. I saw the orange sun through the airplane windows. I saw lots of people in the airport. And, right there and then, I became so excited I almost went berserk. It must have been quite a sight, if anybody noticed, the way I looked as I came off the plane. If you hang around these huge airports at all, you probably see it a thousand times a day: big planes dumping off big bunches of people. Most of them look like they are about to throw up or whatever; then there are likely to be the two or three small, scrawny guys, four at most, hit as hard as I was back then, by whatever it was that hit me, wandering about with such wild eyes you think the best possible thing for them is a dunk in a clear pool. I was excited. You could've offered a trip to the moon and couldn't have gotten me more interested. The arrival—a glorious moment. There were people passing me in all directions, and I stood, confused, tired, ecstatic, and for no good reasons, proud.

As Others Saw Them

The arrival of the ship Great Republic *in San Francisco Harbor in 1869 was described in* Atlantic Monthly*:*

"Her main deck is packed with Chinamen—every foot of space being occupied by them—who are gazing in silent wonder at the new land....

The custom-house officers have done their work here quickly, and perhaps effectually, and now all is ready at the forward gangway. A living stream of the blue-coated men of Asia, bending long bamboo poles across their shoulders, from which depend packages of bedding, matting, clothing, and things of which we know neither the names or the uses, pours down the plank the moment that the word is given, 'All ready!' They appear to be of an average age of twenty-five years—very few being under fifteen, and none apparently over forty years—and though somewhat less in stature than Caucasians, healthy, active, and able-bodied to a man. As they come down unto the wharf, they separate into messes or gangs of ten, twenty or thirty each.... Each man carries his entire earthly possessions, and few are overloaded.... They are all dressed in coarse but clean and new cotton blouses and loose baggy breeches, blue cotton cloth stockings which reach to the knees, and slippers or shoes with heavy wooden soles...and most of them carry one or two broad-brimmed hats of split bamboo and huge palm-leaf fans.... For two mortal hours the blue stream pours down from the steamer upon the wharf; a regiment has landed already, and they still come....

As fast as the groups of coolies have been successively searched, they are turned out of the gates, and hurried away towards the Chinese quarter of the city by agents of the Six Companies."

ANGEL ISLAND

In re Identification of

C H A N S H E E

Wife of Merchant,
applying to enter the
United States.

STATE OF CALIFORNIA, } SS.
ANGEL ISLAND.

Photo of Husband. Photo of Applicant.

LEE YOUK being first duly sworn, deposes and says:

That he is a merchant and a member of the firm of Yan Lung & Co., #30 Waverly Place, San Francisco, California;

That affiant departed from the United States in 1908, from the Port of San Francisco, and has just returned to the United States on the SS "Siberia" of July , 1911, ticket #10148;

That the status of affiant as a boni fide merchant was pre-investigated prior to his departure by the United States Immigration Service and favorably acted upon;

That affiant now brings with him upon his return to America,

Lee Youk was a merchant seeking to return from China with his wife, Chan Shee. Though Lee was permittted to reenter, the document below indicates that his wife's application was denied. The grim records give only a hint of the heartbreak that the Chinese exclusion laws created for families.

Because the Chinese exclusion laws prohibited Chinese laborers and their families, immigrants faced the ordeal of proving that they were eligible for entry. In 1977, Genny Lim and Judy Yung interviewed an old woman named Mrs. Wong, who arrived on Angel Island in 1922. Her husband, a merchant, had immigrated to the United States 10 years before. After establishing a thriving business, he went back to China to bring his wife and their 14-year-old son to the United States. Even so, Mrs. Wong was detained at Angel Island before being permitted to enter the country.

My husband had been in business here for over ten years. There was no trouble getting here. The only thing was that I had to stay on the Island for two weeks or so.

For meals, we went to the big dining hall. At the sound of the bell, we all went down together. All the dishes, including the melon, were all chopped up and thrown together like pig slop. The pork was in big, big chunks. Everything was thrown into a big bowl resembling a wash-tub and left there for you to eat or not. There was cabbage, stewed vegetables, and bits of stewed meat of low quality. Rice was put in another big tub and you served yourself. The Chinese food was not pan-fried, but steamed till it was like a soupy stew. After looking at it you'd lose your appetite.

Sometimes we would receive roast ducks and chickens from here [San Francisco]. But you could only eat a little of it. There was no place to store it, no place to heat it up, so we heated it on top of the radiators for awhile and ate a little of it.

There were twenty of us and two guards. Once a week they took us for a walk around the island. The men were forbidden to go outside. We stayed upstairs and the men stayed downstairs. There were more men than women.

To pass the time away you had to preoccupy yourself any way you could. Some of the ladies who were there for a long time finished a lot of knitting projects. There wasn't a radio or phonograph, not even newspapers or magazines unless you had brought your own. That's why in just two weeks, I was so disgusted and bored of just sitting around! Day in, day out, the same kind of thing.

All we had were rows of bunk beds with a narrow path between the beds, just enough to walk through and not even a chair.... We had a community bathroom, but I never even went to bathe. I kept thinking each day that I would be ready to leave and as each day went by I just waited....

There were very little good times. We never did anything. You mostly sat on your bed and worried, "When am I

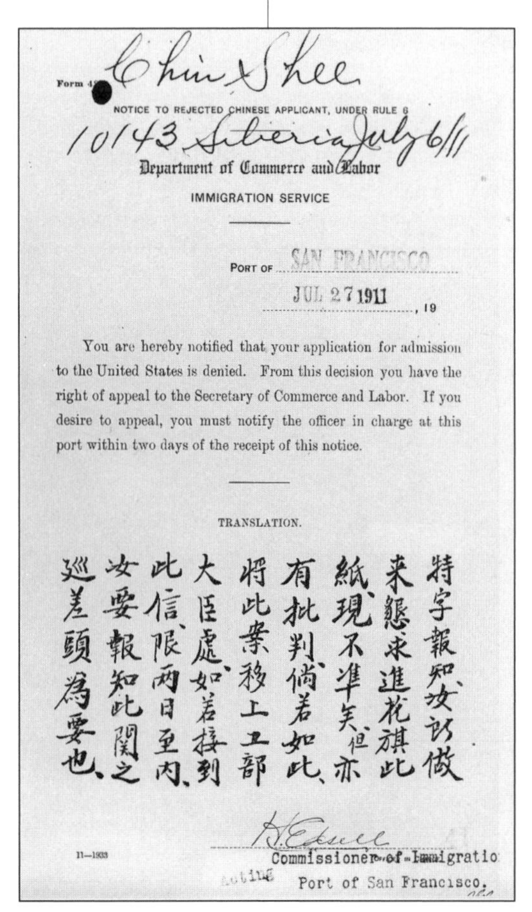

Chin Shee

Form 4[illegible]

NOTICE TO REJECTED CHINESE APPLICANT, UNDER RULE 6

10143 Siberia July 6/11

Department of Commerce and Labor

IMMIGRATION SERVICE

PORT OF SAN FRANCISCO

JUL 27 1911, 19

You are hereby notified that your application for admission to the United States is denied. From this decision you have the right of appeal to the Secretary of Commerce and Labor. If you desire to appeal, you must notify the officer in charge at this port within two days of the receipt of this notice.

TRANSLATION.

特字報知、汝訪做来懇、求進花旗此紙、現不準矣、但亦有批判、倘若如此、将此案移上工部大臣處、如若接到此信、限兩日至內、汝要報知此關之巡差頭為要也、

11—1933

Acting Commissioner of Immigration
Port of San Francisco.

going to get in?" or else, "They're going to deport me!" There wasn't much pleasure under those circumstances.

Others who came through Angel Island also told their stories. Mr. Lee recalled his experience there in 1930, when he was 20 years old.

When we first came, we went to the administration building for the physical examination. The doctor told us to take off everything. Really though, it was humiliating. The Chinese never expose themselves like that. They checked you and checked you. We never got used to that kind of thing—and in front of whites.... If the guard came in and called out a name and said "*sai gaai*" [good luck], it meant that that person was freed to land [leave the island]. If an applicant was to be deported, the guard would make motions as if he were crying.

A group of newcomers on the steps of the Angel Island hospital in 1923. They had crossed the ocean with the hope of finding a new life, and now could only wonder whether they would be allowed to stay.

Mr. Lowe, who was 16 years old in 1939, remembered what it was like while waiting to leave Angel Island.

I had nothing to do there. During the day, we stared at the scenery beyond the barbed wires—the sea and the sky and clouds that were separated from us. Besides listening to the birds outside the fence, we could listen to records and talk to old-timers in the barracks. Some, due to faulty responses during the interrogation and lengthy appeal procedures, had been there for years. They poured out their sorrow unceasingly. Their greatest misery stemmed from the fact that most of them had to borrow money for their trips to America. Some mortgaged their houses; some sold their land; some had to borrow at such high interest rates that their family had to sacrifice. A few committed suicide in the detention barracks.

Women and children waiting in a detention area of Angel Island. Most were probably the wives and children of merchants. Bringing a family into the country was a difficult process, and they often were held on the island for days or weeks before receiving permission to enter the country.

A view of the main buildings at Angel Island. In 1910, an immigration official admitted, "If a private individual had such an establishment he would be arrested by the local health authorities."

THE POEMS OF ANGEL ISLAND

To while away the time at Angel Island, many Chinese wrote poems on the walls of their dormitories. Most of the poems were unsigned—they were simply cries of pain set down by those who believed their efforts to find a better life were in vain. The poems proved to be more valuable than those who wrote them could have guessed. In 1940, the Angel Island detention station was closed. Later, the island was turned into a state park, a gloomy one that few people visited. In 1970, the government planned to destroy the remaining buildings. But a park ranger noticed the inscriptions on the walls and notified Dr. George Araki of San Francisco State University. Araki arranged for the poems to be photographed to preserve them. His efforts inspired the Asian American community to lobby for the preservation of Angel Island. In 1976, the California legislature allotted funds for this purpose.

Him Mark Lai, Genny Lim, and Judy Yung—children of immigrants who passed through Angel Island—collected some of the poems. Here are a few of the writings of those Chinese who wondered if the promise of America would be denied them.

The insects chirp outside the four walls.
The inmates often sigh.
Thinking of affairs back home,
Unconscious tears wet my lapel.

The west wind ruffles my thin gauze clothing.
On the hill sits a tall building with a room of wooden planks.
I wish I could travel on a cloud far away, reunite with my wife and son.
When the moonlight shines on me alone, the nights seem even longer.
At the head of the bed there is wine and my heart is constantly drunk.
There is no flower beneath my pillow and my dreams are not sweet.
To whom can I confide my innermost feelings?
I rely solely on close friends to relieve my loneliness.

America has power, but not justice.
In prison, we were victimized as if we were guilty.
Given no opportunity to explain, it was really brutal.
I bow my head in reflection but there is nothing I can do.

There are tens of thousands of poems composed on these walls.
They are all cries of complaint and sadness.
The day I am rid of this prison and attain success,
I must remember that this chapter once existed.
In my daily needs, I must be frugal.
Needless extravagance leads youth to ruin.
All my compatriots should please be mindful.
Once you have some small gains, return home early.

For what reason must I sit in jail?
It is only because my country is weak and my family poor.
My parents wait at the door but there is no news.
My wife and child wrap themselves in quilt, sighing with loneliness.
Even if my petition is approved and I can enter the country,
When can I return to the Mountains of Tang with a full load?
From ancient times, those who venture out usually become worthless.
How many people ever return from battles?

This is a message to those who live here not to worry excessively.
Instead, you must cast your idle worries to the flowing stream.
Experiencing a little ordeal is not hardship.
Napoleon was once a prisoner on an island.

"The people at Angel Island wrote poems all over the walls, wherever the hand could reach, even in the bathroom.

Some were carved, but most were written with ink. There were many carved in the hall leading to the basketball court, because the wood there was softer. It was not easy finding space on the wall to compose a poem, so sometimes when I thought of something lying in bed, I would bend over and write a poem under my bed, which was made of canvas.... A lot of people there didn't know how to write poetry. They weren't highly educated, but they knew some of the rules of poetry. You can't say the poems were great, but they expressed real feelings."

—Mr. Ng, age 15 in 1931

Many of the poems written by detainees at Angel Island have been preserved for future generations. Today, they are recognized as a cultural treasure that is part of our nation's history.

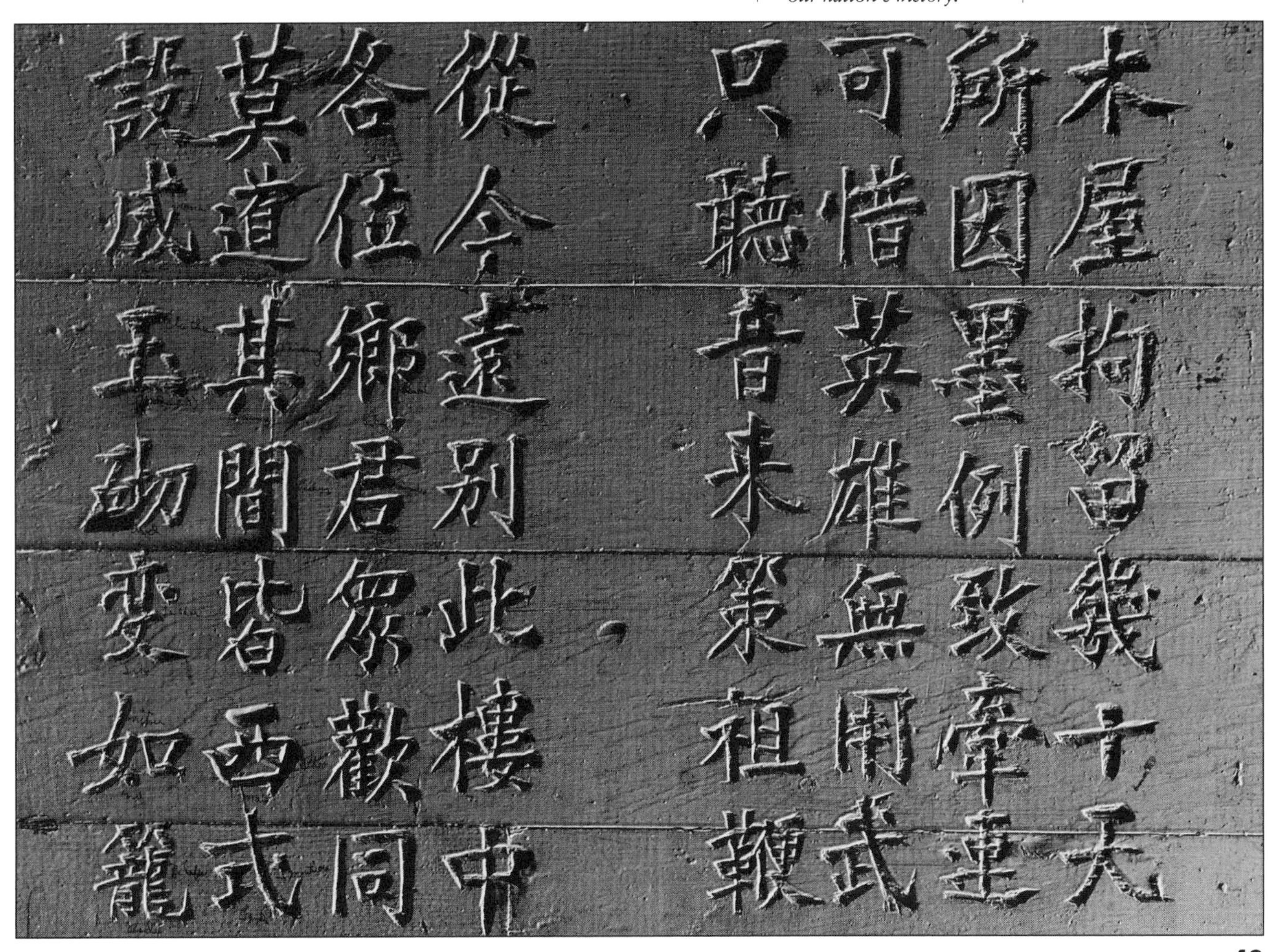

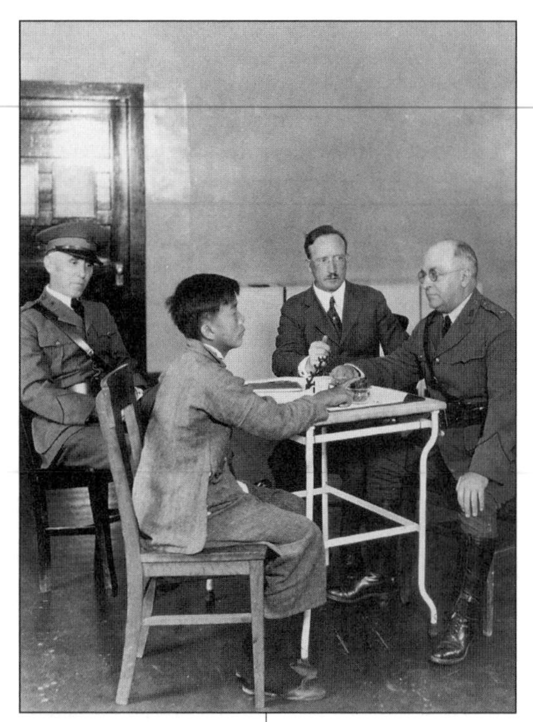

It is easy to imagine the fear and nervousness that this young Chinese felt during hours of intensive questioning about his village and relatives.

THE GRILLING OF A "PAPER SON"

To prepare for their interrogation at Angel Island, "paper sons" purchased coaching books that told the answers they should give. They pored over the books, memorizing the answers that would identify them as residents of the village where their "father" had been born. If several "paper sons" of the same father arrived at the same time, their answers would be compared to see if they matched. Immigrants were questioned several times to see if they gave the same answers to the questions. Here is part of one interrogation.

What is your name?
Leong Sem.

Has your house in China two outside doors?
Yes.

Who lives opposite the small door?
Leon Doo Wui, a farmer in the village; he lives with his wife, no one else.

Describe his wife.
Chin Shee, natural feet.

Didn't that man ever have any children?
No.

How old a man is he?
About thirty.

Who lives in the first house in your row?
Leong Yik Fook, farmer in the village; he lives with his wife, no one else.

How many houses in your row?
Two.

Who lives in the first house, first row from the head?
Yik Haw, I don't know what clan he belongs to.

Why don't you know what clan he belongs to?
I never heard his family name.

Do you expect us to believe that you lived in that village if you don't know the clan names of the people living there?
He never told us his family name.

How long has he lived in the village?
For a long time.

Who lives in the first house, third row?
Leong Yik Gai; he is away somewhere; he has a wife, one son and a daughter living in that house.

According to your testimony today there are only five houses in the village and yesterday you said there were nine.
There are nine houses.

Where are the other four?
There is Doo Chin's house, first house, sixth row.

What is the occupation of Leong Doo Chin?
He has no occupation; he has a wife, no children.

Describe his wife.
Ng Shee, bound feet.

Who is another of those four families you haven't mentioned?
Leong Doo Sin.

Where is his house?
First house, fourth row.

There are two [other] families, who are they and where do they live?
Chin Yick Dun, fifth row, third house.

What is his occupation?
No occupation.

What family has he?
He has a wife and a son; his wife is Chin Shee, natural feet.

Did you ever hear of a man of the Chin family marrying a Chin family woman? [This was forbidden by Chinese custom.]
I made a mistake; her husband is Leong Yick Don.

What is the name and age of that son?
Leong Yick Gai; his house is first house, fourth row.

You have already put Leong Doo Sin in the fourth row, first house.
His house is first house, third row.

You have already put Leong Yick Gai first house, third row.
I am mixed up.

Wong Yow remembered the ordeal of being a "paper son":

My father often talked of wanting to bring his son to the United States. He searched until he found a merchant named Mr. Wong who would sponsor me. My father paid that man $1,650. The merchant, my "paper father," arranged for me to come with another man who was my "paper brother." I was given a document listing me as the merchant's son, and declaring that I was entering the United States as a student. In 1921 I arrived at Angel Island from Hong Kong on the ship *Nile*....

Two events stand out in my memory of life on Angel Island. There was a Chinese interpreter called "Gwongtauh Lou" (baldheaded man) whom all the interviewees feared. Everyone agreed he was mean. I also remember I heard that there were two "brothers" in camp. The officials knew that they had had a banquet at home before they came to the United States, and that a chicken had been slaughtered for the feast. An officer asked each brother separately the color of the chicken's feathers. The younger brother answered yellow, the older black. Based on their conflicting testimonies, the authorities proved that the two were not brothers and deported them.

I remember the ship in Hong Kong. I was the only Chinese girl. A girl! All the old-timers who'd been back and forth to the United States, they said my father was a stupid man. Whoever want a daughter to come to the United States. They always bring sons, they don't bring daughters!

—Helen Wong, on her trip in the 1930s

"Paper sons" often used prompt books like this one to memorize all the details of the village where they were supposed to have been born.

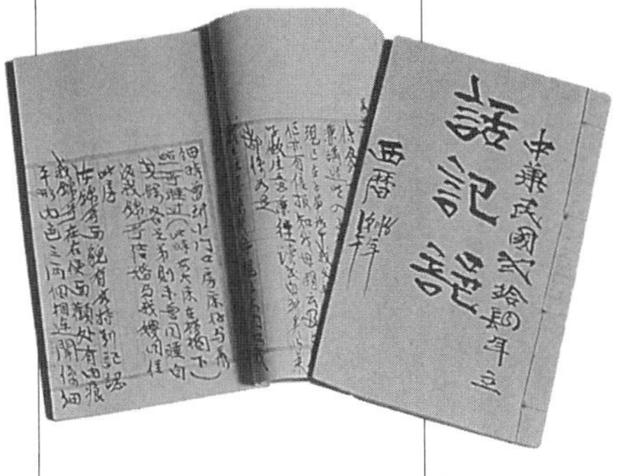

The files of the National Archives of the United States contain many anonymous photographs like this one. They are records of a time when Chinese immigrants were treated virtually as criminals upon their arrival.

"EVERYBODY... HAD A SPLIT PERSONALITY"

The ban on Chinese immigration did not extend to merchants or to students coming to the United States for an education. Thus, some Chinese laborers obtained student certificates as a way of entering the country. But that could create additional problems. Hay Ming Lee's father obtained a false birth certificate after the San Francisco earthquake. He returned to China, intending to bring his son back to the United States. However, he thought it was safer to return as a student. As Hay Ming Lee recalled in the 1970s:

Now my father...when he came in the second time...got a certificate saying he was a student. But that didn't make sense at all. He thought he was so smart being a student, but then, if you come in as a student, how could you bring a son into this country? If he had used his birth certificate, I could have come in as a native son. Instead, we had to go back to the same old thing, "paper son." They had to send me over not as my own father's son, but as the son of another cousin from our village.

"Paper sons" who entered the country illegally lived in fear of being found out. Some paper sons continued to use the names they gave at the immigration station rather than their real names. Even those who had been born here had to be careful not to betray their elders. Pardee Lowe, the son of a prosperous merchant, recalled:

As I grew older I became fully aware of Chinatown's Dr. Jekyll and Mr. Hyde system. It made life for me very complicated. Everybody, officially at least, had a split personality. Aliases were the rule rather than the exception. Father warned me to be extremely careful. For instance, I was not to call Father's business neighbor Mr. Wang in front of any Barbarian, but to hail him as Mr. Fan instead. But in our own Chinatown circle I was to continue to call him Mr. Wang. He wouldn't like it at all, Father said, if I made a mistake.

The fear that Pardee Lowe's father felt was fully justified, as Sam Sue, who was born in Mississippi, explains. Sam's father had entered the United States as a "paper son" in the 1930s. In 1957, Congress passed a law establishing a "confession" program for those who had entered the country by fraud. Supposedly, those who confessed would not be deported if they had a spouse, parent, or child who was a U.S. citizen. However, Sam Sue's father and elder brother found that the program was a trap.

My brother went to Ole Miss (University of Mississippi), and at one point, he was the first Chinese on campus invited to join an all-white fraternity. He was also in the ROTC. Actually it wasn't many years after that that they took away my father's and my oldest brother's citizenship. It was ironic—here he was teaching American government. He was about as American as you can get, and it sort of opened his eyes. Being denaturalized meant he was deportable, so he and my dad had to get waivers, and reapply for citizenship—doing the test again. So they had to be naturalized twice.

There was a confession period for those who came into the country illegally. Many Chinese confessed, and things were okay. But what bugs me, is my dad confessed, and he was nailed to the wall. He came into the U.S. illegally in the 1930s. Later on, he brought my mother and my oldest brother from China. The government took away his citizenship by virtue of him coming here on false papers. He was denaturalized in 1965. That meant what the government gave, it could take away. I mean, Sue is not my real family name. I think it is Jiu.

"Just what is justice in the case of Arthur Lem?" wrote a reporter for the *New York Daily News* in 1961. The federal government had brought criminal charges against Lem, accusing him of entering the country illegally—34 years earlier. At that time, Lem was only 12 years old. Since then, he had lived the American dream, achieving success through hard work. He had built a successful restaurant in Hempstead, Long Island, and had won the respect of his fellow citizens. For serving as chairman of a fund drive for the local YMCA, Lem was elected "Man of the Year" by the business leaders of Hempstead.

Then his life was shattered when a disgruntled employee accused him of being a "paper son." Lem had allegedly helped other Chinese enter the country illegally—including his accuser, who turned Lem in under the government's "confession" program.

At his trial, a number of impressive witnesses testified to Lem's good character. These included judges, a college president, and the Speaker of the New York State Assembly. The jury could not agree on a verdict.

Even so, the federal government insisted on trying Lem a second time. He sold his restaurant and mortgaged his house to pay his legal expenses. Finally, Lem pleaded guilty and received a token sentence. The government had not proved its case, but it had destroyed Arthur Lem's life simply because he shared the dream of all immigrants.

Lee Dick, a Philadelphia merchant, attached this photograph to his application to take his family to China for a visit. Lee stated that he married Lee Chin Farr in China and their child Lee Kim Fong was born in Philadelphia in 1906.

Chinese immigrants were much in demand as cooks in logging camps, mines, and ranches. The art of creating a satisfying meal out of simple ingredients had been developed in China for centuries.

CHAPTER FOUR

A NEW LIFE

The first Chinese in California were welcomed. In January 1852, the governor of the state talked of encouraging "a further immigration and settlement of the Chinese—one of the most worthy classes of our newly adopted citizens." The leading San Francisco newspaper, the *Daily Alta,* said the Chinese "are amongst the most industrious, quiet, patient people among us."

Within a few months, the same California governor was calling the Chinese "coolies" and demanding that the U.S. government pass a law to keep them from working in mines. The *Daily Alta* had changed its tune, too—now calling the Chinese "cunning and deceitful.... They can never become like us and they are not of that race or native character which will ever elevate the social condition of California."

What had happened to cause such a change in opinions? In the governor's campaign for reelection, he found that the white voters resented the success of the Chinese. Since Chinese could not vote, the governor made the obvious political choice. The *Daily Alta* merely reflected the anti-Chinese prejudices of its readers.

None of the anti-Chinese agitators ever accused the Chinese of attacking other people—any violence was directed *against* the Chinese. The great "problem" with the Chinese immigrants was that they worked harder and were more successful than others.

Chinese prospectors became famous for moving into areas that had already been picked over and then finding more gold. The people of Kwangtung Province were familiar with mining techniques. Men from that province had worked in gold mines on the island of Borneo earlier in the century. In California they introduced a device called the Chinese waterwheel, which brought gold out of riverbeds after other miners had given up. Many times, Chinese bought the rights to "worthless" claims and then proved they were still valuable.

The other miners resented the success of the Chinese. A popular song in the gold camps was:

We're working like a swarm of bees scarcely making enough to live
And two hundred thousand Chinese are taking home the gold that we ought to have.

Chinese were frequently chased off their claims. Other miners sometimes beat them or cut off their queues, laughing at "John Chinaman's" fear of losing his hair. (In fact, he could not return to China without his queue—it was against the law for a Chinese male not to have one.)

The laws of California seldom protected the Chinese. In one famous case, a jury convicted a white man named George Hall of killing a Chinese miner. Hall appealed to the California Supreme Court on the grounds that Chinese witnesses had testified against him. In 1854, the court ruled that state law prohibited Indians from testifying against whites—and the Chinese were clearly Indians! Hall was freed.

Yet the Chinese proved their value as workers, and when the California gold fields played out, they soon found other jobs. Some continued as miners when other strikes of silver and gold were discovered in the western states. Big mining companies were well aware of the skills of the Chinese and eagerly hired them. When President Ulysses Grant visited the fabulous silver mines of the Comstock Lode in 1879, a Chinese worker guided the President's party through the diggings.

Possibly the greatest contribution that Chinese immigrants made to the United States was their work on the transcontinental railroad. The federal government had

Chinese laborers laid the iron rails that [linked] the United States from coast to coast. As one railroad manager said, "I do not see how we could have done the work...without them."

awarded this mammoth project to two companies. One, the Central Pacific Railroad, was to begin laying tracks in California and move eastward to link up with the tracks of the other company. The Central Pacific's task was awesomely difficult, for it had to cross the steep, snow-covered cliffs of the Sierra Nevada. When the Central Pacific's white laborers floundered, the railroad reluctantly hired Chinese. The small, slightly built Celestials amazed everyone who saw their speed, skill, and bravery in the most dangerous parts of the construction. A reporter from a New York newspaper called them "a great army laying siege to Nature in her strongest citadel. The rugged mountains...swarmed with Celestials, shoveling, wheeling, carting, drilling and blasting rocks and earth."

The Chinese workers earned such a reputation that afterward they were hired to work on new railroads from Texas to Alaska, and as far east as Tennessee. Chinese laborers laid the iron rails that spread across the country, uniting the United States from coast to coast. As one railroad manager said, "I do not see how we could have done the work...without them."

Other Chinese, taking up their old occupations as farmers and fishermen, proved equally valuable to their adopted country. As pioneers in the agricultural and fishing industries, they enriched California and the other Pacific coast states. Previously, California's farmers had grown mostly wheat. The Chinese cleared the land and planted new crops—grapes, tobacco, hops, sugar beets, strawberries, raspberries, and blackberries. In years to come, these would become part of the fabulous harvest that California now exports to the rest of the nation and the world. Using their knowledge of fruit-drying techniques, the Chinese began the raisin industry around the city of Fresno.

Chinese street vendors became a familiar sight on the West Coast and Hawaii. This "kow kow man" sold hot tea to workers in the Hawaiian sugar fields.

Moving up the Pacific coast, the Chinese also harvested the sea, as the Tanka boat-dwellers had done in Kwangtung for centuries. Canneries were built so the fish could be sent to other parts of the country. Chinese workers took over each part of the process—catching the fish, making the cans by pounding thin strips of metal over an iron cylinder, processing the catch, and then sealing it into cans. The work was hard and the wages low, helping the industry to become profitable. Typically, one worker would can 1,600 fish in a single day, for which he or she was paid $1.50. When a machine was invented to replace much of the human labor, it was called an "Iron Chink." The term *Chink* was a slur against the very people who made the industry possible.

Indeed, despite their many contributions to their new country, the Chinese still faced vicious prejudice. It was greatest in the western states, where the majority of Chinese Americans lived. Racism, of course, was not confined to the West, but the high visibility of the Chinese there made them a target.

The prejudice reached its height after an economic crisis in the early 1870s. Completion of the railroad left thousands of laborers without jobs. Several of the richest silver mines played out at this time, adding miners to the list of unemployed. A drought caused crop failures in 1876, and farm laborers flocked to the cities in search of work.

Unfortunately, the tide of Chinese immigration reached its height at the same time. Though Chinese still made up only 10 percent of the population of California, the *New*

On September 2, 1885, the townspeople of Rock Springs, Wyoming, attacked a nearby settlement of Chinese coal miners. Twenty-eight Chinese were killed, their huts burned, and their bodies mutilated.

York Times warned that the western states were becoming "Chinese colonies." In 1871, mobs attacked a Chinese section of Los Angeles, killing 22 people.

Unemployed white workers resented the fact that the hardworking Chinese, who were willing to work for lower wages, could find jobs. Labor organizers raised the old cry that the Chinese were "coolies." In San Francisco, a rabble-rouser named Denis Kearney attracted a following with his cry: "The Chinese must go! They are stealing our jobs!" In July 1877, three days of rioting broke out in San Francisco. Dozens of Chinese businesses and homes were burned.

Various organizations, such as the Workingman's Party and the League of Deliverance, were formed to give the anti-Chinese movement greater influence. Their members organized boycotts of goods made by Chinese Americans and staged demonstrations to force factories and mills to dismiss their Chinese workers.

Republicans and Democrats, seeking votes from the western states, both adopted anti-Chinese statements in their platforms for the 1880 Presidential election. The Democrats declared that "their influence being corrupt and corrupting, Chinese should not have lot or part among the American people."

In 1882, in response to the Chinese "threat," Congress overwhelmingly passed the Chinese Exclusion Act. At that time, Chinese made up only 0.2 percent of the population of the United States.

Even these few Chinese were not safe from violence and prejudice. No one knows how many were killed or chased away from farms, fisheries, and factories. Additional restrictive laws were passed by both the state and federal governments, forbidding Chinese to own land or marry whites. Indeed, if a white female American citizen married a Chinese "permanent resident," she automatically lost her U.S. citizenship. The American Federation of Labor, led by an English immigrant, Samuel Gompers, continually urged the passage of more legislation to keep Chinese from working in trades and taking jobs away from white Americans.

Virtually without legal protection, the Chinese became the prey of bandits and lawless mobs. On September 2, 1885, the townspeople of Rock Springs, Wyoming, attacked a nearby settlement of Chinese coal miners. Twenty-eight Chinese were killed, their huts burned, and their bodies mutilated.

The Chinese in Hawaii fared somewhat better than those on the mainland. Chinese were already working in Hawaii in 1835, when a New Englander arrived on the island of Kauai to establish a sugar plantation. Feeling that Chinese were better workers than the native Hawaiians, American businessmen were soon bringing Chinese contract laborers to the islands. After the United States annexed Hawaii in 1898, the Chinese Exclusion Act applied there, and the sugar planters turned to other Asians, such as Japanese, Koreans, and Filipinos, who were not barred from entering the country. The proportion of Chinese in the Hawaiian population dropped steadily, as it did in the mainland United States.

For the next 60 years, Chinese Americans saw themselves depicted in newspapers, magazines, and movies as mysterious, sinister people. At best, they were "the inscrutable Chinese," who concealed their true feelings behind a mask of politeness.

For centuries, Chinese doctors have known about the healing powers of natural herbs. This man in Wyoming was one of many who found a demand for his services.

As Others Saw Them

"[T]hey soon flocked into the mining regions in swarms, well-satisfied to work over the old abandoned claims left and deserted by others. They were welcomed by the mining communities with open arms, as it was soon discovered that the Chinese would not preempt or locate any new mining grounds, desiring only to buy at a fair price the old worked-out claims, which had been abandoned... and well did these disciples of Confucius merit the title of scavengers of the mining regions, for many of the old claims which had been abandoned as worthless, were not so in fact, as it was soon discovered that from many of them the Chinese were taking out large amounts of gold."

—An observer of Chinese miners in 1890

Scenes like this one, showing Chinese and non-Chinese prospectors together, were rare. Other miners often resented the Chinese for their ability to find gold where others had failed.

THE GOLD RUSH

The gold rush drew large numbers of Chinese to America. Unfortunately, no one has discovered any account written by a Chinese forty-niner in search of gold. Modern Chinese American writers have tried to imagine what the experience was like. In China Men, *Maxine Hong Kingston wrote of her father, a teacher in a Chinese village, listening to the stories of old men, sojourners who had been to Gum Shan and returned.*

The Gold Mountain Sojourners were talking about plausible events less than a century old. Heroes were sitting right there in the room and telling what creatures they met on the road, what customs the non-Chinese follow, what topsy-turvy land formations and weather determine the crops on the other side of the world, which they had seen with their own eyes. Nuggets cobbled the streets in California, the loose stones to be had for the stooping over and picking them up. Four Sojourners whom somebody had actually met in Hong Kong had returned from the Gold Mountain in 1850 with three thousand or four thousand American gold dollars each. These four men verified that gold rocks knobbed the rivers; the very dirt was atwinkle with gold dust. In their hunger the men forgot that the gold streets had not been there when they'd gone to look for themselves.

C. Y. Lee's novel, Land of the Golden Mountain, *describes the adventures of a 17-year-old Chinese girl who disguises herself as a boy to go to the California gold fields. Though her real name is Mai Mai, she is known to the other Chinese in her group as Straw Sandal. The following passage describes the thrill of her first "strike."*

At the camp she picked a pail and a shovel and rushed to the mines in the hills. She found Ta Ming's mine. He was shoveling dirt out of a four-foot deep trench beside a few others, their heads popping in and out, perspiration flying. Longevity [a boy Mai Mai met on the ship coming over] was digging too. Everybody seemed to have caught gold fever. Mai Mai leaped into the trench and plunged into work.

The sun poured down on her. Presently perspiration ran down her face and her hands blistered. But she kept hacking the dirt away feverishly. Then she shoveled the dirt into her pail as though every bit of it were pure gold.

"Throw the soft dirt away," Ta Ming shouted at her. "There is no gold in the top layer."

"Yes, sir," Mai Mai answered in English, forgetting that she was talking to Ta Ming. She tossed the dirt out like the others and went deeper until she hit a layer of coarse sand. She filled her pail with it and dashed to the river.

At the river she discovered she didn't have a pan. She dashed back and borrowed one. Ta Ming watched her, shaking his head, half amused and half worried.

Mai Mai panned her dirt breathlessly, her heart pounding. She rocked and twisted the submerged pan until the mud was washed away. When she brought it out of the water and examined the contents, there was nothing left but coarse sand and some pebbles. Disappointed, she threw it away, dashed back to the trench and resumed her digging like a tireless beaver. Ta Ming came over to her, still shaking his head.

"Go deeper," he advised. "Dig four feet, at least. You can only find gold in the blueish clay. Here, go wash this." He dumped beside her a pail of the heavier dirt.

Mai Mai had no time to talk. She grabbed it and made a dash for the river. Just as she was about to reach the water she tripped over a rock and scattered the contents of her pail wide and far. In spite of a painful toe and bruised knees and hands, she scrambled to her feet and raced back to the trench.

"Go easy, Straw Sandal," Ta Ming said, more worried than amused now. "The gold is here. It is not going to run away."

Mai Mai filled the pail with the dirt and once more dashed back to the river, her clothes wet with perspiration and her pigtail flying. She panned the dirt excitedly, her eyes strained over the shallow water, searching for the first glow of gold. The moment she discovered a tiny fleck among the black sand in her pan she let out a cry that startled Longevity, who was panning beside her.

"What's the matter, Straw Sandal?" Longevity asked. "Bitten by a crab?"

"Gold!"

Longevity looked and made a face. "You have enough in that pan to fill a small fingernail."

"How much have you panned?"

"Enough to fill ten large fingernails, at least."

"Then I have no time to talk," she said, racing back for more pay dirt. She forgot the heat, the painful toe and the bruises. She worked feverishly until after dusk and collected a small sack of black sand with some gold flecks in it. When Mr. Quon came to collect the gold she asked him if she could keep it for a night. She wanted to sleep on it for luck. Mr. Quon looked at the meager contents of the sack and said with a shrug, "Enough to buy a toothpick. Keep it."

The remark didn't cure her of gold fever. Back at the tent she could not take her eyes off her gold. She fingered it, squeezed it and bounced it on her palm. "*Wei,* Straw Sandal," Four-eyed Dog said, "why don't you smell it and chew it? That's the only thing you haven't done so far."

"A story was told of a Chinese mining company headman named Ah Sam, who, for twenty-five dollars, bought a dirt-floored log cabin from six American miners. As Ah Sam had suspected, the previous owners of the cabin had been careless and had permitted gold dust to fall all over the dirt floor. He put his crew to work washing the dirt in the cabin--and reportedly realized $300 worth of gold dust from his original $25 investment."

—Daniel amd Samuel Chu, Passage to the Golden Gate

The hope that one could gain a fortune in a short time drew thousands of people to California from all over the world. Even in this multinational crowd, the Chinese stood out with their distinctive clothing, headgear, and queues.

Carrying his equipment on his back, this man was a "placer" miner. He would shovel sand from the bottom of a stream into his wooden cradle, rocking it back and forth to wash away the sand, while the heavier flakes of gold sank to the bottom.

These Chinese American firefighters (above) won first prize in the Fourth of July hose team contest in the frontier town of Deadwood, South Dakota, in 1888.

Chinese Americans gathered abalone along the Pacific coast (right). The meat was a delicacy in China, and the pearly shells could be sold for making jewelry.

Using fine-mesh nets imported from China, Chinese American fishermen took shrimp, squid, and fish from the waters off the California coast. The catch was sold fresh in San Francisco markets or dried for export to China.

PIONEERS

"Life on a plantation is much like life in a prison," one Chinese sugarcane worker in Hawaii remembered. The field hands had to wear bangos—brass disks with an identification number—on a chain around their necks. Overseers did not bother to learn the men's names; they addressed them by number.

Some Chinese who settled in the United States took up their old occupations as fishermen and farmers. They laid the foundations for the fishing and produce industries of California. A white journalist described a California Chinese fishing village in 1854.

Many of our readers may not be aware that on the south side of Rincon Point, San Francisco, near the mouth of Mission Creek, there is a settlement of Chinese well worth a visit. It consists of about one hundred and fifty inhabitants, who are chiefly engaged in fishing. They have twenty-five boats, some of which may be seen at all hours moving over the waters—some going to, others returning from the fishing-grounds. The houses are placed in a line on each side of the one street of the village, and look neat and comfortable. Here and there, a group is seen making fish-lines, and with their rude machines, stacking in heaps the quantities of fish which, lying on all sides around, dry in the sun, and emit an ancient and fish-like odour. The fish which they catch consist of sturgeon, rates, and shark, and large quantities of herring. The latter are dried whole, while the larger are cut into thin pieces. When they are sufficiently dry, they are packed in barrels, boxes, or sacks, and sent into town to be disposed of to those of their countrymen who are going to the mines or are bound upon long voyages. An intelligent Chinaman told us that the average yield of their fishing a day was about three thousand pounds, and that they found ready sale for them at five dollars the hundred pounds.

Chinese also worked in fish canneries up the Pacific coast from California to Alaska. By many accounts, the work was too difficult and the pay too low to attract white workers. Chinese-owned agencies in California hired teams of men and sent them north with an English-speaking Chinese foreman. They had to work long hours; they slept in a crowded, barnlike structure called the China house; and they were paid about $165 per season. They sent much of their pay back to relatives in China. Mont Hawthorne, who built canneries in Alaska, reported:

After working around Chinaboys for close to 15 years, I got to see their side of things pretty near as good as I could see my own. I used to think it was wrong for them fellows to send their money back to the old country. But the first boy I ever talked to about it said: "I come here to work so I can send money to my father and mother. They have a big family in China and only two acres of ground. If I do not work here and send them money, they will starve."

All they expected to get out of life was hard work and the

Chinese workers on the San Francisco docks loading rice onto a ship bound for Alaska. The date on the photo is April 12, 1906, six days before an earthquake devastated San Francisco.

Polly Bemis

Because the overwhelming majority of early Chinese immigrants were male, prostitutes were much in demand. Few women took up the occupation voluntarily. It was common for young women and girls to be brought from China as slaves. Though contracts sometimes bound them for a specific number of years, few were able to win their freedom at the end of that period. One who did—Polly Bemis—became a pioneer on the western frontier.

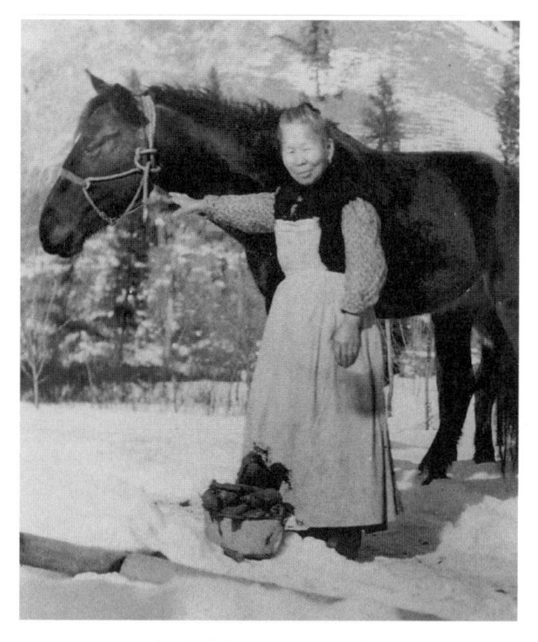

Her amazing life began in northern China in 1853, where she was named Lalu Nathoy. She was sold to bandits for two bags of seed and eventually was shipped to the United States. In 1872, a Chinese saloonkeeper in an Idaho mining camp purchased Lalu, giving her the name Polly. Some time later, a white miner named Charlie Bemis won her in a poker game.

Life was brutal and dangerous in the rough mining town. In a gunfight, Charlie Bemis's cheekbone was shattered by a bullet. Polly dug the fragments out of his cheek with a crochet hook and used her knowledge of herbal medicine to heal his wound. Charlie married her and the two of them settled down on a 20-acre plot of land on the Salmon River. She saved his life a second time by dragging him unconscious from their burning cabin.

Because there was no doctor in the area, homesteaders sought Polly's help for illnesses and wounds. By the time she died in 1933, she was one of the most respected citizens of Grangeville, Idaho. The townspeople named the stream running through her property Polly Creek in her memory.

promise that when they died their bones would be shipped back to the old country to be buried in the graves of their ancestors. And even that wasn't done until they was buried over here long enough for the flesh to waste away. It saved expense and shipping space to dig them up, pack each man's bones in a little metal box, and send them back home in a sizable consignment.

Polish writer Henryk Sienkiewicz described the Chinese truck gardens that encircled 19th-century San Francisco.

Whoever goes to the outskirts of the city will perceive at the ends of the unfinished streets, on the hills, valleys, and slopes, on the roadsides, in fact, everywhere, small vegetable gardens encircling the city with one belt of greenness. The ant-like labor of the Chinese has transformed the sterile sand into the most fertile black earth.... The fruits and vegetables, raspberries, and strawberries under the care of Chinese gardeners grow to a fabulous size. I have seen strawberries as large as small pears and heads of cabbage four times the size of European heads, and pumpkins the size of our wash tubs.

In the early 1970s, Johnny Kan remembered his childhood, half a century earlier, when he peddled vegetables in a valley town in the foothills of the Sierra Nevada.

We had a very difficult time growing up. At one time there had been over two thousand Chinese living in Grass Valley and working in the gold mines. But by the time we got up there it had dwindled down to a few hundred. A lot of the men had fallen back on vegetable peddling. I remember when we were very young, after school my sister and I would go from house to house peddling strawberries for something like a nickel a basket.... My mother was a good baker, and she would send us down to Chinatown on Sunday mornings with covered baskets to sell *cha shew bau* [steamed meat dumplings] to a few Chinese who would come down from the hill, the gold mines, or wherever they were living.

Around 1855, Chinese tended their vegetable gardens in San Francisco. The site of this photo is Union and Pierce Streets, today a fashionable area of the modern city.

The image of the western frontier created by American movies ignored the presence of Chinese Americans such as this cowboy in Nevada.

Lue Gim Gong

Lue Gim Gong came to America in 1872 at the age of 12. He worked in a shoe factory, earning a little more than a dollar a day, but he looked forward to getting rich. That dream was never realized; instead, Lue Gim Gong would enrich his adopted country far more than it would ever repay him.

He arrived at a bad time, when the United States was suffering from a recession. Factories reduced production and cut their workers' pay. The Chinese immigrants' willingness to work for low wages enabled them to keep their jobs but spurred anti-Chinese protests by labor agitators.

Factory owners in the eastern and southern states began to import Chinese workers from the West Coast. Lue Gim Gong volunteered to go with a group of his countrymen to a shoe factory in North Adams, Massachusetts. Knowing little or no English, the Chinese were bewildered by the hostility they encountered. The factory had brought them to break a strike of its regular workers.

Eventually, the dispute was settled, but some of the Chinese stayed on. The residents of North Adams set up a school to teach them English. Here, Lue Gim Gong befriended Fanny Burlingame, the daughter of the American diplomat who had negotiated the Burlingame Treaty in 1868, which allowed Chinese citizens to immigrate to the United States and become permanent residents but not citizens. She hired him to work in her garden, where he amazed her by breeding a new strain of tomato. He had learned the skill of plant pollination from his mother in China.

Fanny Burlingame moved to Florida and sent Lue money to join her. He was pleased to find some orange trees growing there. Two centuries earlier, Spanish colonists had brought the fruit from China, but the trees had never adapted well to Florida's wet climate.

Lue began to crossbreed the Florida oranges with varieties from other countries. Eventually, he produced a tree that bore fruit nearly all year round. A nursery signed a contract with Fanny Burlingame to market this amazing new variety. It became the standard orange that made Florida one of the nation's leading fruit producers. A few years later, Lue Gim Gong's orange won the Wilder Medal, given to honor important new plant varieties. In a rare exception to the anti-Chinese immigration laws, he was allowed to become a citizen of the United States.

In 1903, Fanny Burlingame died, leaving Lue her orange groves. Sadly, the gentle Chinese horticulturist was not a clever businessman. Others cheated him out of the royalties from his plant discoveries. In his old age, he limped through his garden, growing barely enough food to stay alive. He died in 1925, virtually penniless.

"A person passing on the SPCRR [South Pacific Coast Railroad] train the other day said he counted thirty-two graves by the wayside of the Chinamen that were killed at the recent explosion in tunnel No. 3."
—*Santa Cruz Sentinel,* November 1879

THE RAILROAD BUILDERS

When Congress approved funds for building the transcontinental railroad, it awarded the contracts to two companies. One, the Central Pacific, was to move east from California and eventually link up with the Union Pacific, which was heading west from the Mississippi River. Congress turned the construction project into a race by rewarding the companies for each mile of track they built. But the difficulties of the western route soon caused the Central Pacific to fall behind schedule.

Even so, when Charles Crocker urged his partners in the Central Pacific to hire Chinese laborers, they were reluctant. Governor Leland Stanford, one of the other owners, had promised in his election campaign to exclude Chinese immigrants from the state. But the Chinese soon made the owners into believers. They worked in gangs of 12 to 20 men; wages were $30 to $35 a month per man, the same as for white laborers. The Chinese demanded that their own food be brought by wagon from San Francisco. Their meals included oysters, abalone, cuttlefish, dried bamboo sprouts and mushrooms, noodles, rice, dried seaweed, sweet rice crackers, and dried fruit. The Chinese insisted that kettles of hot tea be kept in their camps at all times. This diet, especially the boiled tea, probably protected them from some of the diseases that afflicted white workers, who drank unboiled water. As Stanford himself admitted, "Without them [the Chinese], it would be impossible to complete the western portion of this great enterprise."

None of the Chinese workers who completed the superhuman task of building the transcontinental railroad wrote their own stories. In China Men, *Maxine Hong Kingston honored her grandfather, Ah Goong, by imagining what his work on the transcontinental railroad must have been like.*

A Chinese tea-carrier brings the freshly brewed drink that the Chinese workers insisted on having. Boiling the water to make tea saved the Chinese from the diseases that afflicted other workers.

When cliffs...ended the road, the workers filled the ravines or built bridges over them. They climbed above the site for tunnel or bridge and lowered one another down in wicker baskets made stronger by the lucky words they had painted on four sides. Ah Goong got to be a basketman because he was thin and light. Some basketmen were fifteen-year-old boys. He rode the basket barefoot, so his boots, the kind to stomp snakes with, would not break through the bottom. The basket swung and twirled, and he saw the world sweep underneath him; it was fun in a way, a cold new feeling of doing what had never been done before....

Swinging near the cliff, Ah Goong stood up and grabbed it by a twig. He dug holes, then inserted gunpowder and fuses. He worked neither too fast nor too slow, keeping even with the others. The basketmen signaled one another to light the fuses. He struck match after match and dropped the burnt matches

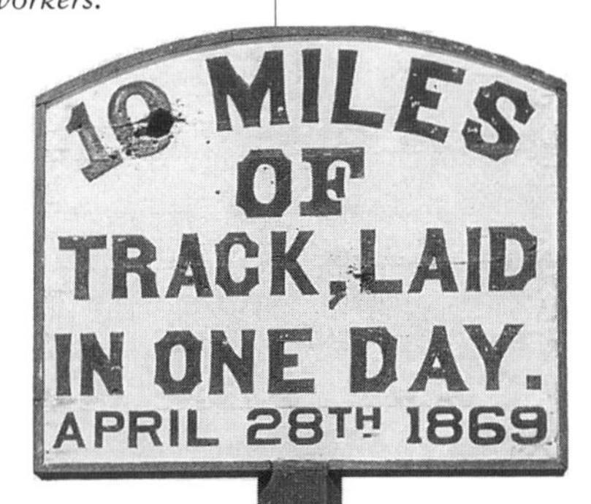

After a group of Irish laborers on the rival Union Pacific Railroad set a record for laying eight miles of track in a single day, Charles Crocker of the Central Pacific declared that his Chinese workers could do better. They did—an achievement that this sign commemorated.

over the sides. At last his fuse caught; he waved, and the men above pulled hand over hand hauling him up, pulleys creaking.... "Hurry, hurry," he said. Some impatient men clambered up their ropes. Ah Goong ran up the ledge road they'd cleared and watched the explosions....

This time two men were blown up. One knocked out or killed by the explosion fell silently, the other screaming, his arms and legs struggling. A desire shot out of Ah Goong for an arm long enough to reach down and catch them.... His hands gripped the ropes, and it was difficult to let go and get on with the work. "It can't happen twice in a row," the basketmen said the next trip down. "Our chances are very good. The trip after an accident is probably the safest one." They raced to their favorite basket, checked and double-checked the four ropes, yanked the strings, tested the pulleys, oiled them, reminded the pulleymen about the signals, and entered the sky again....

Then it was autumn, and the wind blew so fiercely the men had to postpone the basketwork.... The food convoys from San Francisco brought tents to replace the ones that whipped away. The baskets...carried cowboy jackets, long underwear, Levi pants, boots, earmuffs, leather gloves, flannel shirts, coats. They sewed rabbit fur and deerskin into the linings. They tied the wide brims of their cowboy hats over their ears with mufflers....

The days became nights when the crews tunneled inside the mountain, which sheltered them from the wind, but also hid the light and sky. Ah Goong pickaxed the mountain, the dirt filling his nostrils through a cowboy bandanna.... Beneath the soil, they hit granite. Ah Goong struck it with his pickax, and it

Charles Crocker's partners protested that Chinese were too weak to endure the hardships of constructing a railroad over the high mountains of the Sierra Nevada. Crocker replied that "any race that could build the Great Wall of China could build a railroad."

Sixty-two miles east of Sacramento, the Chinese reached the foothills of the Sierra Nevada. They filled in valleys like this one to provide a base for the trestle that would carry the trains. Working only with picks, shovels, and hand-pulled carts, the Chinese won the respect of all who saw them.

As Others Saw Them

"They [the Chinese] were a great army laying siege to Nature in her strongest citadel. The rugged mountains looked like stupendous anthills. They swarmed with Celestials shoveling, wheeling, carting, drilling, and blasting rocks and earth."

—*Albert Richardson, a correspondent for the* New York Tribune

The Chinese workers proved so valuable that before long the Central Pacific sent agents to China to recruit more. In 1865, Leland Stanford reported, "We have assurances from leading Chinese merchants that...the Company... will be able to procure during the next year not less than 15,000 laborers."

jarred his bones, chattered his teeth. He swung his sledgehammer against it, and the impact rang in the dome of his skull. The mountain that was millions of years old was locked against them and was not to be broken into....

When the foreman measured at the end of twenty-four hours of pounding, the rock had given a foot. The hammering went on day and night. The men worked eight hours on and eight hours off. They worked on all eighteen tunnels at once. While Ah Goong slept, he could hear the sledgehammers of other men working in the earth. The steady banging reminded him of holidays and harvests; falling asleep, he heard the women chopping mincemeat and the millstones striking....

One day he came out of the tunnel to find the mountains white, the evergreens and bare trees decorated, white tree sculptures and lace bushes everywhere. The men from snow country called the icicles "ice chopsticks."... The dynamiting loosed blizzards on the men. Ears and toes fell off. Fingers stuck to the cold silver rails. Snowblind men stumbled about with bandannas over their eyes.... Ah Goong looked at his gang and thought, If there is an avalanche, these are the people I'll be trapped with, and wondered which ones would share food....

The men who died slowly enough to say last words said, "Don't leave me frozen under the snow. Send my body home. Burn it and put the ashes in a tin can. Take the bone jar when you come down the mountain."...

Spring did come, and when the snow melted, it revealed the

past year, what had happened, what they had done, where they had worked, the lost tools, the thawing bodies, some standing with tools in hand, the bright rails. "Remember Uncle Long Winded Leong?" "Remember Strong Back Wong?" "Remember Lee Brother?" "And Fong Uncle?" They lost count of the number dead; there is no record of how many died building the railroad. Or maybe it was demons doing the counting and chinamen not worth counting. Whether it was good luck or bad luck, the dead were buried or cairned next to the last section of track they had worked on. "May his ghost not have to toil," they said over graves....

He spent the rest of his time on the railroad laying and bending and hammering the ties and rails. The second day the China Men cheered was when the engine from the West and the one from the East rolled toward one another and touched. The transcontinental railroad was finished. They Yippee'd like madmen. The white demon officials gave speeches. "The Greatest Feat of the Nineteenth Century," they said. "The Greatest Feat in the History of Mankind," they said. "Only Americans could have done it," they said, which is true. Even if Ah Goong had not spent half his gold on Citizenship Papers, he was an American for having built the railroad. A white demon in top hat tap-tapped on the gold spike, and pulled it back out. Then one China Man held the real spike, the steel one, and another hammered it in.

The Chinese railroad workers acquired such a reputation that they were in demand for other projects in the west and south. This group was photographed near Monterey, California, in 1889.

Maxine Hong Kingston

"What I am doing...is claiming America," Maxine Hong Kingston said in 1980. In her first two books, *The Woman Warrior* (1976) and *China Men* (1980), Kingston combined the story of her own family with legends, folklore, and history to produce an American epic. Kingston gave a voice to the Chinese Americans who toiled on the railroads and in the sugarcane fields and laundries. By "claiming America," she asserted the rightful place of Chinese Americans in the history of their adopted land.

Kingston's father, Tom Hong, came to this country in 1924. Trained to be a mandarin, or scholar-official, he could find no place to use his skills. Calling himself Tom after the inventor Thomas Edison, he took a job in a laundry and saved enough money to bring his wife here. Their first child, Ting Ting (later Maxine), was born in Stockton, California, in 1940.

Ting Ting's mother told her six children stories that had been passed down in her village, Sze Yup, for generations. Her "talk-stories" of ghosts and ancestors, heroes and heroines, tragedies, romances, and triumphs were stored in Ting Ting's memory. Later, they would become the basis of her great books.

At school, Ting Ting never raised her hand in class, for her parents spoke only Chinese at home, and her English was clumsy. By the time she was nine, however, she was writing little poems in English and calling herself Maxine. She earned straight As and won a scholarship to the University of California at Berkeley. She graduated in 1962 and married Earll Kingston the same year. The couple moved to Hawaii in 1967.

Nine years later, Kingston's first book, *The Woman Warrior*, brought her instant fame. The title refers to Fa Mu Lan, the sword-wielding heroine of a Chinese legend. Kingston identified with her struggle to avenge the downtrodden and oppressed. In the course of the book, Kingston describes her own battle for independence from her domineering mother. The sequel, *China Men*, takes the reader into the time when the first Chinese came to the United States. It describes how the "China men," like her own father, fought to find a place for themselves in America.

In 1862, a committee of the California State Legislature observed that it had received a list of 88 Chinese who had been killed by Americans of European descent. Eleven of these murders had been committed by collectors of the Foreign Miner's Tax. The report of the committee concluded: "It is a well known fact that there has been a wholesale system of wrong and outrage practiced upon the Chinese population of this state, which would disgrace the most barbarous nation upon earth."

"THE CHINESE MUST GO"

From the time they first arrived, Chinese endured taunts and the sting of prejudice. Johnny Kan, who earned his living selling vegetables in a mining camp, remembered:

A particular character whose name was Duck Egg, he walked with a stoop. He was getting old and living a little bit off charity. You know, whenever you walked into a Chinese store or laundry, or any kind of business, if you were Chinese you would receive hospitality, food, and whatever help they can give you. Duck Egg was one of those.... You know, it was a rough bunch of characters who came to California to the Mother Lode, and they would treat Chinese as inferior and stone them because we looked different and acted different. So they used to stone the vegetable peddlers, they would stone Duck Egg, and they used to throw stones at us on the way home. Oh, they would wrap rocks in snowballs and throw them at us. Of course, children we must forgive, because they were innocent, but at that time it was a very confusing thing for me.

The state of California passed a number of laws aimed at the Chinese. Here is a partial listing:

1853: Foreign Miner's Tax forbade any noncitizen of the United States from taking gold from the mines of the state "unless he shall have a license."

1855: Immigrant Tax was designed "to discourage the immigration to this state of persons who cannot become citizens thereof."

1879: New California constitution declared, "The presence of foreigners ineligible to become citizens of the United States is...dangerous to the well-being of the State, and the Legislature shall discourage their immigration by all the means within its power." The constitution prohibited anyone except native-born whites, foreigners of the white race, and those of African descent from owning or inheriting land.

1880: Act Relating to Fishing in the Waters of This State was passed by the California legislature to prohibit all aliens "from fishing, or taking any fish, lobsters, shrimps, or shell-fish of any kind for the purpose of selling or giving to another person to sell."

The city of San Francisco also passed restrictive laws:

1870: San Francisco "Cubic Air" Ordinance required every building in which people lived to contain "at least five hundred

The Wasp *was a 19th-century magazine that carried inflammatory articles against the Chinese immigrants. This cartoon from an 1876 issue warned that New York's Statue of Liberty would have a Chinese figure if it appeared in "our harbor"—San Francisco Bay.*

cubic feet for each adult person." Most buildings housing Chinese violated the law, and so many Chinese were arrested for violating the law that the jails themselves also violated.

1876: San Francisco Queue Ordinance required the sheriff to cut the hair of every prisoner sentenced to jail, within one inch of the prisoner's scalp. A Chinese prisoner, Ho Ah-Kow, brought a lawsuit against Sheriff Matthew Nunan, who had cut his hair. Three years later, in 1879, the United States Circuit Court decided that the ordinance was unconstitutional and that Ho Ah-Kow's civil rights had been violated. The court ordered Nunan to pay Ho Ah-Kow $10,000.

1879: San Francisco Laundry Ordinance set a series of fees for laundries. Any laundry using one horse-drawn vehicle was required to pay a fee of $2 every three months. A laundry using two such vehicles had to pay a $4 fee. But a laundry using *no* such vehicles had to pay a $15 fee. This law was obviously aimed at the Chinese laundries, which used men carrying poles to deliver their laundry.

A 19th-century child's toy gun featured a man kicking a Chinese when the trigger was pulled.

On January 18, 1855, a convention of Shasta County, California, miners passed resolutions against the Chinese, as did many other mining communities at this time. The Shasta Convention resolved:

That the immense number of Chinamen flocking into this country has become an evil too great to be borne. That it becomes the American miners to take prompt and decisive measures to stop an evil that threatens to overwhelm us;

That it is our opinion that no measures short of prohibition and total expulsion of all Chinamen thence will remedy the evil from which we suffer;

That we, the miners of Shasta County, forbid Chinamen from working in the mines of this county after the 25th of February, 1855.

This advertisement for a washing-machine company promised that its invention would drive Chinese laundry workers out of San Francisco and back to China.

The American Federation of Labor supported the political movement that resulted in the Chinese Exclusion Act. Samuel Gompers, president of the AFL, later declared:

"There is a great deal of difference between a Chinaman and an American. I am not in favor of excluding the Chinaman from our shores because he is a Chinaman, but because his ideas and his civilization are absolutely opposed to the ideals and civilization of the American people. Never in the history of the world have the Chinese been admitted into a land save to dominate or to be driven out."

Denis Kearney, an Irish immigrant, was the most famous of the rabble-rousers who whipped up prejudice against the Chinese in the 1870s. At that time, Huie Kin was working on a farm near Oakland. He saw firsthand the frightening results of the bigotry.

The sudden change of public sentiment towards our people in those days was an interesting illustration of mob psychology.... The useful and steady Chinese worker became overnight the mysterious Chinaman, an object of unknown dread. When I landed [in 1868], the trouble was already brewing, but the climax did not come until 1876–1877. I understand that several causes contributed to the anti-Chinese riots. It was a period of general economic depression in the Western States, brought about by drought, crop failures, and reduced output of the gold mines, and on the top of it came a presidential campaign.... There were long processions at night, with big torchlights and lanterns, carrying the slogan "The Chinese Must Go," and mass meetings where fiery-tongues flayed the Chinese bogey. Those were the days of Denis Kearney and his fellow agitators, known as sandlot orators, on account of their vehement denouncements in open-air meetings. To Kearney was attributed the statement which showed to what extremes political demagogues could go: "There is no means left to clear the Chinamen but to swing them into eternity by their own queues, for there is no rope long enough in all America wherewith to strangle four hundred millions of Chinamen." The Chinese were in a pitiable condition in those days. We were simply terrified; we kept indoors after dark for fear of being shot in the back. Children spit upon us as we passed by and called us rats. However, there was one consolation: the people who employed us never turned against us, and we went on quietly with our work until the public frenzy subsided.

An ad for rat poison played on the popular prejudice that Chinese ate rats.

A cartoon from the 1880s shows a sinister, many-armed Chinese doing a variety of jobs, holding a bag of savings labled "For China." The artist's message was that less-than-human Chinese took jobs from "real" Americans who would spend their wages here. In fact, most immigrants of all groups sent money to families in their homelands.

Some Chinese openly protested against their treatment. A man named Kwang Chang Ling wrote a series of remarkable letters to the San Francisco Argonaut *in 1878. Kwang reminded his readers that Chinese had invented such things as the compass, gunpowder, and printing long before they were known in European nations. Kwang minced no words in replying to the bigots who argued that Chinese were inferior to other people.*

You demand every privilege for Americans in China but you would deny the same privileges to Chinamen in America, because in your opinion the presence of the Chinese among you is a menace to your civilization. You shrink from contact with us, not because you regard us as mentally or bodily inferior, for neither fact nor argument will support you here—but rather because our religious code appears to be different from yours, and because we are deemed to be more abstemious in food, clothing, and shelter....

Let me...correct one great misapprehension in respect to the Chinese. You are continually objecting to his morality. Your travelers say he is depraved; your missioners call him ungodly; your commissioners call him uncleanly.... Yet your housewives permit him to wait upon them at table; they admit him to their bed-chambers; they confide to him their garments and jewels; and even trust their lives to him by awarding him supreme control over their kitchens and the preparation of their food. There is a glaring contradiction here....

The slender fare of rice and the other economical habits of the peasant class [of China], which are so objectionable to your lower orders and the demagogues who trumpet their clamors, are not the result of choice to Chinamen; they follow poverty. The hard-working, patient servants that you have about you today, love good fare as well as other men, but they are engaged in a work far higher than the gratification of self-indulgence; they are working to liberate their parents in China...and so long as their labor continues to strike off the fetters from their beloved ones will they continue to practice their noble self-abnegation. When this emancipation is complete, you will find the Chinaman as prone as any human creature to fill his belly and cover his back with good things.

Another educated Chinese, Chung Sun, sought—in vain—to remind Americans of the noble principles on which the United States was founded. His letter, translated from Chinese by an American friend, appeared in the newspaper of Watsonville, California, where Chung worked as a ditch digger for $1.50 a day.

Unlearned as you may think us to be, we are not wholly ignorant of your history.... We are taught to believe that...your government is founded and conducted upon principles of pure justice and that all of every clime, race, and creed are here surely protected in person, liberty, and property....

I left the loved and ever venerated land of my nativity to seek [in the United States] that freedom and security which I

The Exclusion Act caused verbal protests in China as well as a boycott of U.S. goods. The newspaper Nih Nih Sing (Fukien Daily News) *in May 1905 commented:*

"In order to exclude the Chinese, the United States adopted force, disregarded justice, ignored humanity and violated international treaties. This was a great insult imposed upon all of us four hundred million Chinese. The reason they dared to do this while we silently accepted the result was because they were united, but we were not; they were strong but we were weak...

The immigration issue is not the Cantonese issue but the Chinese issue. The insult degrades not the Cantonese alone but the Chinese as well. If we four hundred million Chinese do not exert ourselves to support the boycott, we are not a nation any more.... On the contrary, if we support the Cantonese to fight to end the unjust exclusion policy and our government takes measures complying with the public opinion, then the United States, no matter how strong it is, will be forced to moderate its exclusion policy."

Denis Kearney, one of the leading anti-Chinese rabble-rousers, was himself an immigrant—from Ireland. A cartoonist ridiculed the tendency of other groups to make the Chinese a scapegoat for their problems.

Anna May Wong

In American motion pictures until the 1950s, white actors usually took the roles of Chinese characters, whether they were villains or heroes. Boris Karloff played the sinister Fu Manchu, and Werner Oland portrayed the clever Chinese-American detective Charlie Chan. In the first half century of American movies, virtually the only real Chinese star was Anna May Wong.

She was born in Los Angeles's Chinatown in 1907. Her parents, who owned a laundry, named her Liu Tsong. During her childhood, the brand-new movie industry grew up in Los Angeles's most famous section, Hollywood. Often, the movie makers recruited extras for crowd scenes right off the street, and that was how Liu Tsong made her screen debut, at the age of 12. Though her father was bitterly opposed to her working as an actress, she pursued her dream.

Four years later, she landed her first major role, as a Mongol slave in *The Thief of Bagdad*, which starred Douglas Fairbanks. As with many other immigrants who became movie actors, her real name was regarded as "too foreign" for a screen credit. A movie executive dubbed her Anna May Wong, which was a more "suitable" mix of Chinese and American.

For nearly four decades, until her last movie in 1960, Anna May Wong was the first person movie producers thought of when they wanted to fill the part of a female Asian. She played Chinese, Japanese, even an Eskimo—to American audiences, all "Orientals" looked much the same.

Sadly, the parts she played were stereotypes as well. A movie critic once referred to "the bitchy Oriental dragon lady out of an old Anna May Wong movie." Wong's roles usually required her only to look sinister and mysterious, for that was virtually the only kind of Asian portrayed in the movies of her day.

When silent movies gave way to "talkies," Anna May Wong had to assume an accent that sounded "Oriental." In real life, she spoke with the same accent as other Americans who had been born and raised in the United States.

could never hope to realize on my own, and now after some months' residence in your great country, with the experience of travel, study and observation, I hope you will pardon me for expressing a painful disappointment. The ill treatment of [my] own countrymen may perhaps be excused on the grounds of race, color, language and religion, but such prejudice can only prevail among the ignorant....

Being a man of education and culture I am capable of other work than digging in the streets, but my philosophy teaches me, any *useful* work is more honorable than idleness. I shall therefore, with patience and resignation continue to dig with an abiding hope for something better.... I shall try to be charitable as well as just to all mankind, but as a people will hardly correct their faults without knowing them, I write this in a spirit of kindness, notwithstanding my ill treatment, and ask you to publish it.

Many times, mobs attacked Chinese settlements, both in cities and rural areas. On October 24, 1871, a mob of several hundred whites shot, hanged, and stabbed 19 Chinese to death in Los Angeles. One of the victims was a Chinese doctor named Gene Tung, who offered to pay his captors to spare his life. They hanged him anyway, stole his money, and cut off one of his fingers to obtain a ring. Meanwhile, the mob ransacked Chinese stores and homes, stealing everything of value.

Some of the anti-Chinese violence was prompted by labor organizers. Chinese coal miners working at Rock Springs, Wyoming Territory, were suddenly attacked by white miners on September 2, 1885. Twenty-eight Chinese were killed and 15 wounded. Their settlement was looted and then burned, causing more than $147,000 in damages. No one was ever punished for the outrage. The Chinese miners later sent an account of the massacre to the Chinese consul in New York.

Up to the time of the recent troubles we had worked along with the white men, and had not the least ill feeling against them. The officers of the companies employing us treated us and the white man kindly, placing both races on the same footing and paying the same wages.

Several times we had been approached by the white men and requested to join them in asking the companies for an increase in the wages of all, both Chinese and white men. We inquired of them what we should do if the companies refused to grant an increase. They answered that if the companies would not increase our wages we should all strike, then the companies would be obliged to increase our wages. To this we dissented....

On the morning of September 2, a little past seven o'clock, more than ten white men, some in ordinary dress and others in mining suits, ran into Coal Pit No. 6, loudly declaring that the Chinese should not be permitted to work there. The Chinese present reasoned with them in a few words, but were attacked with murderous weapons, and three of their number wounded. The white foreman of the coal pit, hearing of the disturbance, ordered all to stop work for the time being....

About two o'clock in the afternoon a mob, divided into two gangs, came toward "Chinatown," one gang coming by way of the plank bridge, and the other by way of the railroad bridge.... After two Chinese were killed [by gunfire] the Chinese now, to save their lives, fled in confusion in every direction.... Whenever the mob met a Chinese they stopped him and, pointing a weapon at him, asked him if he had any revolver, and then approaching him they searched his person, robbing him of his watch or any gold or silver.... Some of the rioters would let a Chinese go after depriving him of all his gold and silver, while another Chinese would be beaten with the butt ends of the weapons before being let go. Some of the rioters, when they could not stop a Chinese, would shoot him dead on the spot, and then search and rob him.... Some, who took no part either in beating or robbing the Chinese, stood by, shouting loudly and laughing and clapping their hands.

There was a gang of women that stood at the "Chinatown" end of the plank bridge and cheered; among the women, two of them each fired successive shots at the Chinese....

The Chinese who were the first to flee...were scattered far and near, high and low, in about one hundred places. Some were...lying hid on the grass, or stooping down on the low ground. Every one of them was praying to Heaven or groaning with pain. They had...seen the whites, male and female, old and young, searching houses for money, household effects, or gold, which were carried across to "Whitemen's Town."

Some of the rioters...set fire to the Chinese houses. Between 4:00 P.M. and a little past 9:00 P.M. all the camp houses belonging to the coal company and the Chinese huts had been burned down completely.

An officer in the U.S. Army drew this picture of the Rock Springs massacre. Afterward, some Chinese returned to Rock Springs, with federal troops assigned to guard them.

A Death in Detroit

In June 1982, a young Chinese American named Vincent Chin joined some friends in a neighborhood bar in Detroit, Michigan. Two white automobile workers named Michael Nitz and Ronald Ebens, mistaking Chin for a Japanese, began to make insulting remarks. They were resentful about competition from the Japanese auto industry, which was putting Americans out of work. After a fight broke out, Chin left the bar. Nitz and Ebens got a baseball bat from their car and pursued him. While Nitz held Chin's arms, Ebens shattered his skull with the bat and killed him.

The city prosecutor allowed Nitz and Ebens to plead guilty to manslaughter. They were sentenced to three years' probation and fined $3,780 each. Not only in Detroit, but all over the United States, Chinese Americans were outraged and shocked. Most agreed with Lily Chin, Vincent's mother, who said, "This happened because my son is Chinese. If two Chinese killed a white person, they must go to jail, maybe for their whole lives."

Vincent Chin's great-great-grandfather had worked on the transcontinental railroad. His father had served in the U.S. Army during World War II. Both of his parents had worked in a laundry and hoped their son could have a better career. An architecture student, Vincent was working as a draftsman when he was murdered.

Asian Americans saw the Vincent Chin case as a sign that they were still not fully accepted. They pointed to school curricula that failed to include the contributions of Asians to this country. "What disturbs me," said George Wong of the Asian American Federation of Union Membership, "is that the two men who brutally clubbed Vincent Chin to death in Detroit in 1982 were thinking the same thoughts as the lynch mob in San Francisco's Chinatown one hundred years ago: 'Kill the foreigners to save our jobs! The Chinese must go.'"

Memorial service for Vincent Chin.

A Chinese couple in Arizona in the late 19th century

CHAPTER FIVE

PUTTING DOWN ROOTS

From the time of their arrival in San Francisco, the Chinese found security and comfort by living in their own section of the city. Like many other immigrants, they wanted a place where they could hear the familiar language of their homeland, eat food prepared Chinese-style, and follow the customs and traditions they had learned as children. The San Francisco city directory of 1852-53 listed nine Chinese businesses on Sacramento Street. This became the heart of what is still called Chinatown.

As anti-Chinese prejudice grew, Chinatown—not only in San Francisco, but also in other cities, both large and small—became a refuge from violence and prejudice. There was safety in numbers. An invisible wall separated the Chinatowns from the outside world; within that boundary, the Chinese Americans created their own society.

Chinatown shops sold medicines, food, and other articles imported from China. Chinese stores were social centers that often doubled as banks. People could obtain loans there or arrange to have the money they earned sent home to their families. Public bulletin boards featured announcements of public interest. The first Chinese American newspaper, the *Golden Hills News,* began in San Francisco in 1854. Many others followed.

Small Confucian and Buddhist temples could be found tucked away on side streets, providing another link with the Chinese homeland. These religious centers were sometimes called "joss houses," because of the Chinese practice of burning joss sticks, or incense, as a sign of devotion.

During the first century of Chinese immigration, the proportion of males to females was very high. In 1900, about 95 percent of Chinese Americans were male. The members of this "bachelor society" often shared dormitory-like rooms that offered little more than a place to sleep. The bachelors' favorite recreation was gambling. The Chinatowns had many gambling halls, marked by signs (in Chinese) reading "Riches and Plenty" or "Winning Hall." Lotteries and such games as fan-tan and mah-jongg were popular.

Chinatown was not entirely a male society, however. Some merchants did bring their wives and children. But these Chinese American families led sheltered lives until well into the 20th century. Most public schools would not accept nonwhite children. Chinese wives were carefully protected, for a Chinese woman walking alone was regarded as a prostitute and might even be kidnapped.

From the beginning, the Chinese Americans formed organizations for mutual protection and assistance. Some were modeled after similar ones in China. Family, or clan, associations included anyone who shared the same family name. In the United States, it was natural for members of the same clan to rely on each other for support and friendship. The clan associations became a permanent part of Chinese American life.

Chinese Americans also formed district associations. These were composed of people from the same geographical area—those who spoke a familiar dialect and shared memories of their home villages.

Seeing the need for unity, the leaders of six district associations agreed in 1860 to form a larger group to represent the interests of all Chinese in the United States. Officially named the Chinese Consolidated Benevolent Association (CCBA), the group was popularly known as the Six Companies.

The CCBA became the most powerful and respected Chinese organization, with branches in every city that had a sizable Chinese population. It organized opposition to discriminatory laws, both state and federal. As early as 1884, the San Francisco chapter of the CCBA opened a school for Chinese children. The CCBA's leaders eventually served as the government of Chinatown, and non-Chinese city officials recognized it as such.

However, the CCBA's powers were so great that some modern Chinese American scholars have called it "despotic." The CCBA required Chinese businessmen to pay membership dues and also collected fees for issuing "exit permits" to Chinese who wished to return home. It persuaded the steamship companies not to sell tickets to any Chinese without a CCBA permit.

Other organizations, called *tongs,* formed a second power center within Chinatown communities. The tongs were modeled after secret societies in China whose avowed purpose was to overthrow the ruling dynasty and restore a Chinese emperor to the throne.

In the United States, unfortunately, many tongs became deeply involved in criminal activities. Some brought "sing-song girls" from China illegally. Though their headquarters bore innocent-sounding names such as the "Hall of Victorious Union" and the "Chamber of High Justice," most tong members were gangsters who terrorized other Chinese. Ruthless thugs, known as highbinders, shook down merchants for "protection" money.

Tong wars sometimes broke out when one tong encroached on the illegal business of another. In 1875, the first of these wars began when the Suey Sing Tong posted a red-paper *chun hung,* or challenge, on a San Francisco Chinatown bulletin board. The Suey Sing's rival, the Kwong Duk Tong, responded with a notice of its own. The resulting battle, fought with hatchets and knives, left several men dead and the Suey Sing victorious. For the next half century, scenes like this were repeated in Chinatowns in other cities. Mainstream newspapers printed sensational stories about the tong wars, adding to the prejudice against all Chinese Americans. Though the last tong war ended in 1927, the tongs retain some influence even today.

In 1900, more than 95 percent of all Chinese Americans were males. Some married non-Chinese women, including Mexicans and Filipinos, and raised families that were part of the multiethnic "melting pot."

In 1904, some California-born Chinese, discontented with the CCBA's failure to control the tongs and to overturn the ban on Chinese immigration, started a new organization called the Native Sons of the Golden State. Its name, quite intentionally, was an echo of the Native Sons of the Golden West, one of the anti-Chinese groups that wanted to allow immigration only for whites. By taking this name, the American-born Chinese proclaimed that they, too, were citizens and true Americans. When the organization spread to other states, it became known as the Chinese American Citizens Alliance. It continues to be a leader in fighting anti-Chinese prejudice.

For decades, the most visible Chinese businesses were laundries and restaurants. An unsuccessful gold miner named Wah Lee is said to have opened the first Chinese American laundry. In 1851, the sign "Wash'ng and Iron'ng" appeared over his door at the corner of Grant Avenue and Washington Street in San Francisco. No one knows what made him choose that way of making a living, for in China, men did not do laundry. But in the gold-rush boomtown, where fortunes were made overnight, people would pay almost any price for food, clothing, and services. It was common for people to ship their dirty laundry to Honolulu or even to Hong Kong, paying as much as a dollar to have a single shirt cleaned and ironed. Wah Lee saw a need and filled it; he charged $5 for a dozen shirts.

The Chinese laundry exploded on the American scene. By one estimate, there were more than 2,000 Chinese laundries in San Francisco by 1870. Countless others opened in Chicago, New York, and virtually anywhere Chinese Americans went. As one Chinese laundryman said, "I don't know how the laundry became a Chinese enterprise in this country. But I think they just learned it from each other.... All one has to do is watch how others do it."

Similarly, Chinese restaurants became part of the American scene almost as soon as the first immigrants arrived. In San Francisco, triangular flags of yellow silk hung outside Chinese restaurants, and gold miners soon learned that they were assurance of a good meal inside. One miner wrote that "the best eating houses in San Francisco are those kept by Celestials and conducted Chinese fashion. The dishes are mostly curries, hashes and fricasees served up in small dishes and as they are exceedingly palatable, I was not curious enough to enquire as to the ingredients."

Chinese restaurants have maintained their popularity ever since. Today, Chinese takeout food, delivered hot in small white boxes, is an American custom. Chinese vegetables and spices, as well as frozen Chinese food, are available in supermarkets across the United States.

The sojourners, as well as their American-born descendants, retained close ties to China and political events there. Chinese Americans provided important support to groups plotting the overthrow of the Manchu dynasty. Sun Yat-sen, one rebel leader, formed a "Revive China Society" in Honolulu in 1894. The group's first members were Canton-born Chinese who were living in Hawaii.

Ten years later, after his first attempts to organize a revolt had failed, Sun came to San Francisco. Learning of his arrival, the Chinese consul asked the U.S. government to deport him to China, where he faced a death sentence. From the customhouse, Sun smuggled a note to the editor of the *Chung Sai Yat Po,* a Chinatown newspaper. The editor hired a lawyer to obtain Sun's release.

The *Chung Sai Yat Po* printed pamphlets describing the revolutionary struggle and sent them to tongs throughout the United States. Sun made a tour of 27 American cities, speaking and raising money for his cause. To attract a crowd for the speech, a Chinese opera was presented beforehand. Chinese Americans responded generously, for they believed that the weakness of their home country was one reason why they faced such bigotry overseas. Sun Yat-sen, who wore Western-style suits and cut off his queue, represented the modern leadership that Chinese Americans believed China needed.

On October 10, 1911, Sun was in Colorado when he received news that the long-awaited revolution had at last broken out. As he rushed back to begin organizing a new government, Chinese Americans celebrated the downfall of the Manchu dynasty (also called the Ching dynasty) by cutting off their queues.

China's struggles were far from over, but Sun Yat-sen would remain a Chinese hero. Sun's picture and the flag of his political party, the Kuomintang, occupied places of honor in Chinese American communities.

During the 1930s, when Japan invaded China, Chinese Americans rallied to the defense of their ancestral land. They staged demonstrations urging the U.S. government to stop sending scrap iron and oil to Japan.

In 1941, the United States's attitude toward China changed. Late that year, the United States entered World War II after Japan bombed the American naval base at Pearl Harbor in Hawaii. Now, the United States and China became allies in the great worldwide conflict. Chinese Americans wore buttons proclaiming, "I'm not Jap." Native-born Chinese Americans signed up to join the U.S. Army and Navy in large numbers.

It was embarrassing to the U.S. government to ban immigrants from a wartime ally. President Franklin Roosevelt urged Congress to repeal the Chinese Exclusion Act. Roosevelt said, "We must be big enough to acknowledge our mistakes of the past and to correct them." In 1943, the U.S. Congress repealed the Chinese Exclusion Act. After more than 60 years, Chinese immigrants were once more welcome in America.

Chung Sai Yat Po *was one of four Chinese-language newspapers in San Francisco in the early 20th century. This car proudly carried the flag of the new Republic of China in a parade.*

A fortune-teller carries on his trade in the streets of San Francisco's Chinatown around 1900. He probably used the Yi Jing, *one of the ancient books supposedly edited by Confucius himself.*

THE FIRST CHINATOWN

Almost as soon as Chinese came to America, they banded together in a distinctive community. In 1851, a San Francisco newspaper reported, "About 200 Chinamen have...formed an encampment on a vacant lot at the head of Clay Street. They have put up about 30 tents, which look clean and around which are scattered the various articles of Chinawares and tools which they brought over. They look cheerful and happy." This was the beginning of what was first dubbed "Little China," and later, "Chinatown." Because there was not enough timber to supply the demands of the gold-rush boomtown, entire buildings were actually shipped from China and reassembled in San Francisco. Huie Kin, arriving in 1868, found:

San Francisco's Chinatown was made up of stores catering to the Chinese only.... Our people were all in their native costume, with queues down their backs, and kept their stores just as they would do in China, with the entire street front open and groceries and vegetables overflowing on the sidewalks. Forty thousand Chinese were then resident in the bay region, and so these stores did a flourishing business.

By the 1870s, Chinatown had definite boundaries—Kearney, Stockton, Sacramento, and Pacific Streets. It could not expand farther because it was hemmed in by the business district of the

Herbal medicine shops were part of every Chinese American community. Chinese pharmacists and doctors used such drugs as quinine, digitalis, and ephedrine long before their value was known to Western doctors.

city, and conditions became ever more crowded. The Chinatown businesses were small, as one writer noted: "A space in a wall of no greater dimensions than a large dry goods box furnishes ample room for a cigar stand; and a cobbler will mend your shoes in an area window, or an unused door step. Nothing goes to waste."

San Francisco's white citizens regarded Chinatown as a blight, and the city health officer declared, "Some disease of a malignant form may break out among [the Chinese] and communicate itself to our Caucasian population. Their mode of living is the most abject in which it is possible for human beings to exist."

To outsiders, Chinatown became a feared and mysterious place, "a system of alleys and passages...into which the sunlight never enters; where it is dark and dismal even at noonday. A stranger attempting to explore them, would be speedily and hopelessly lost.... Often they have no exit—terminating in a foul court, a dead wall, a gambling or opium den."

Then came disaster. According to one who lived through it:

The great earthquake came and everything was destroyed. It was in April 1906. And it was around five o'clock in the morning. We were all in bed.... I wake up, and here everything is shaking. And then, it shook quite a while, and of course I was only a child, but, then, here went everything tumbling down! And then, I went out to the door and looked out way down California Street...and the street had a big hole there. And then later on the sparks, the fire, the blaze began to start down around Montgomery Street.

The earthquake and fire destroyed most of San Francisco's Chinatown, and proposals were made to force the Chinese to locate elsewhere. However, while the city government debated the issue, the Chinese residents quickly rebuilt their houses and shops in the same place. Chinatown was there to stay.

Fengshui and Wells, Fargo

In 1852, businessman John Parrott prepared to construct San Francisco's first stone building. He imported granite blocks from China, each one marked with Chinese characters indicating how they should fit together. But the Chinese laborers hired by Parrott shook their heads when they saw the site, on the corner of Montgomery and California Streets. The place had bad *fengshui*, they told Parrott.

In China, *fengshui* ("wind-water") experts are consulted to determine the proper location of all buildings, so that the gods of nature will not be offended. Parrott did not believe in *fengshui* and ordered the construction to proceed. Two major tenants, a bank and an overland shipping company, moved in.

The California Chinese steered clear of the spot. They took their gold to the express agents across the street—Wells, Fargo & Co. Wells, Fargo made a determined effort to win Chinese customers by advertising in their newspapers and issuing business directories in Chinese.

In 1855, a financial panic caused the bank in the Parrott Building to fail. The shipping company there also collapsed. Wells, Fargo weathered the crisis, in part because its Chinese customers remained faithful.

When Wells, Fargo wanted to move to larger offices, it asked the Chinese if the Parrott Building could be made safe. The Chinese burned incense and made an offering of rice and tea to the gods and declared that now the building had proper *fengshui*. From its new headquarters, Wells, Fargo became one of the most prosperous banks in the western states.

This man's queue indicates that this photo was probably taken before 1911, when the Manchu dynasty was overthrown. After that time, many Chinese American men cut off their long queues as a sign of support for the new Chinese republic.

Illiterate men who wished to send a message to their families paid public letter writers, who sometimes practiced their trade in the street.

A well-to-do father and son (right) stroll through New York's Chinatown in the 1930s.

A Chinese shop owner in Honolulu.

CHINATOWNS U.S.A.

As anti-Chinese bigotry raged in the West, Chinese sought safety by moving eastward, often using the very railroads they had built. Chinatowns grew up in cities and towns, ranging in size from a handful of people to the tens of thousands in New York, Chicago, and Boston. Chinese felt security in living with others like themselves. As one Chicago Chinatown resident said in the mid-1920s:

Most of us can live a warmer, freer, and a more human life among our relatives and friends than among strangers.... It is only in Chinatown that a Chinese immigrant has society, friends, and relatives who share his dreams and hopes, his hardships and adventures. Here he can tell a joke and make everybody laugh with him; here he may hear folktales told which create the illusion that Chinatown is really China.

Amy Tan, one of the best of today's Chinese American writers, described Chinatown in her novel The Kitchen God's Wife. *Pearl, the central character, has returned to her childhood home for her grand-aunt's funeral and is headed for the flower shop that her mother and aunt own.*

As I turn down Ross Alley, everything around me immediately becomes muted in tone. It is no longer the glaring afternoon sun and noisy Chinatown sidewalks filled with people doing their Saturday grocery shopping....

On the right-hand side of the street is the same old barbershop, run by Al Fook, who I notice still uses electric clippers to shear his customers' sideburns. Across the street are the same trade and family associations, including a place that will send ancestor memorials back to China for a fee. And farther down the street is the shopfront of a fortune-teller. A hand-written sign taped to the window claims to have "the best lucky numbers, the best fortune advice," but the sign taped to the door says: "Out of Business."...

And now I'm at Sam Fook Trading Company, a few doors down from the flower shop. It contains shelves full of good-luck charms and porcelain and wooden statues of lucky gods, hundreds of them. I've called this place the Shop of the Gods ever since I can remember. It also sells the kind of stuff people get for Buddhist funerals—spirit money, paper jewelry, incense, and the like....

And now I come to the flower shop itself. It is the bottom floor of a three-story brick building. The shop is about the size

Joe Shoong

Before the repeal of the Chinese Exclusion Act in the 1940s, some Chinese established successful businesses outside Chinatown. One outstanding example is Joe Shoong, who founded the National Dollar Stores, a chain of clothing stores in the western part of the United States.

Shoong was born in San Francisco in 1879; his parents had emigrated from China around the middle of the 19th century. He opened his first store in Vallejo, California, in 1903. It sold inexpensive dry goods, such as sheets, clothing, and bolts of cloth. Four years later, when San Francisco was rebuilding after the devastating earthquake, Shoong moved his business there. By 1928, he had 16 stores and named the company the National Dollar Stores.

National Dollar Stores were exactly that—no item in the store cost more than a dollar. Though every store had a Chinese manager, the salesclerks were usually white, as were most of the customers. National Dollar Stores became known for their bargain-priced clothing. In 1938, *Time* magazine called Joe Shoong "the richest, best-known Chinese businessman in the U.S."

His success, however, rested on the exploitation of Chinese laborers. Most of the goods were made by low-paid garment workers in Chinatown sweatshops. In 1937, a group of workers organized the Chinese Ladies' Garment Workers and went on strike for higher wages and better working conditions. After 13 weeks, Shoong closed the factory. In that depression year, he could find plenty of willing workers elsewhere.

Nevertheless, Shoong contributed part of his fortune to charity, helping to raise funds for the Chinese Hospital in San Francisco. Before his death in 1961, he established the Joe Shoong Foundation, which grants college scholarships to Chinese American students.

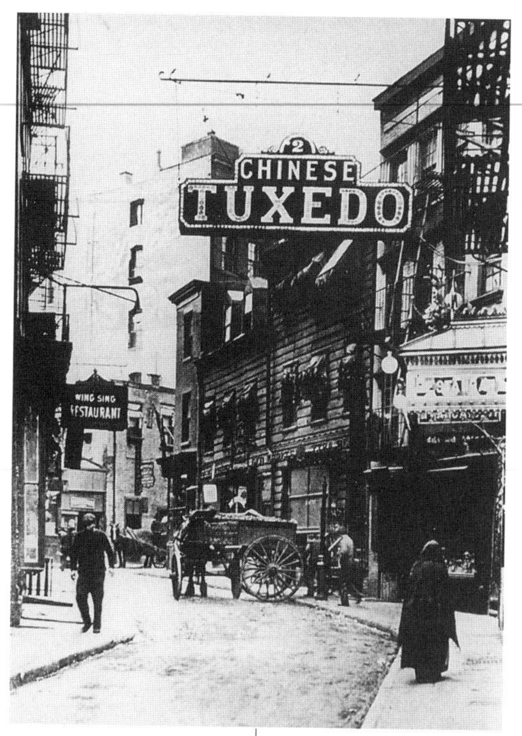

"Mulberry Bend" in New York City's Chinatown, around 1890. Though New York City's newspapers often printed lurid tales of Chinatown's tong wars, non-Chinese New Yorkers knew that the community's restaurants served some of the best food in the city.

of a one-car garage and looks both sad and familiar. The front has a chipped red-bordered door covered with rusted burglar-proof mesh. A plate-glass window says "Ding Ho Flower Shop" in English and Chinese....

The place is dimly lit, with only one fluorescent tube hanging over the cash register—and that's where my mother is, standing on a small footstool so she can see out over the counter, with dime-store reading glasses perched on her nose.

In Chinatown theaters, traveling groups of actors presented the distinctive Chinese-style opera. These plays, accompanied by music that featured cymbals, drums, and flutes, lasted for many hours. In fact, some plays were part of a long saga that went on through many performances, much like a modern soap opera. Because the plays were based on familiar tales, the audiences turned the performances into social occasions. Drifting in throughout the evening, they brought food and drink and chatted with friends and neighbors. The actors saved their most spectacular scenes for late in the evening when the theaters were filled with latecomers.

Eileen Lee remembered the theater in New York:

Going to the Chinese opera at Sun Sing theatre was a real treat for me when I was a little girl. My Mom would come home from work earlier than usual to cook dinner for my grandparents and me. Then after dinner she would pack a snack basket for us. My Mom never came to the opening, she came during the second act because it was cheaper. Anyway, what I remember most was all the color of the costumes they wore, especially the antennas on their headpieces. I thought it was funny the way they moved and bounced. Another thing I remember is my grandmother telling me that the man on stage was really a woman actress. I think I was one of the few kids that liked "pots and pans music" (as my non-Chinese speaking cousins would call it).

Grace Mok had similar memories:

Opera would run four or five nights. Everybody was eating in the theater. And children running up and down the aisles. I used to wonder how are you supposed to enjoy this. Only a Chinese place would do that. I grew up with it, I didn't know the difference. I thought, "Well, children are just part of the show."

This modern-day poster shows that Chinese theater is still popular among the Chinese community in New York.

The Chinese Theatre

The theater at 5-7 Doyers Street in New York's Chinatown also attracted some non-Chinese patrons, who sat in a separate section.

"In 1895, the Chinese Theatre, located at 5-7 Doyers Street, opened its doors to eager Chinese operagoers. White writers and journalists, however, more often visited the theater to search for sinister goings-on. In the early years of this century, stories appearing in the mainstream press often emphasized the theater's location, a place which reporters often called the 'Bloody Angle,' a bend reputed for its tong-related murders. During this period of journalistic sensationalism, the Chinese Theatre became famous more for the murders of 'hatchet men' than for its musical life."

—Mary Lui, curator of "Red Boat on the Canal: Cantonese Operatic Arts in New York Chinatown," Chinatown History Museum

A headdress used in some of the colorful Chinese opera performances.

The script for a Chinese play was usually just an outline for the actors to follow. They improvised their lines and action, giving the familiar plays a fresh performance each time.

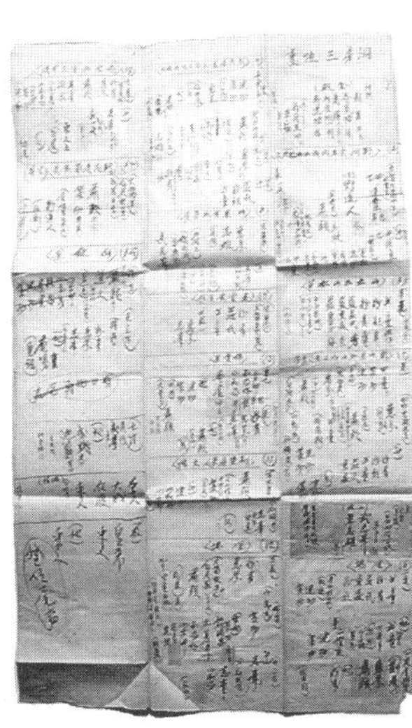

Musicians ordinarily accompanied Chinese opera performances. Percussion and string instruments enhanced the action on stage and played a leading role in the orchestra.

THE BACHELOR SOCIETY

Loneliness and despair marked the lives of many of the Chinese American "bachelors." One of them, living in San Francisco, summed up his existence in 1915 in a poem:

Come to think of it, what can I really say?
Thirty years living in the United States—
Why has life been so miserable and I, so frail?
I suppose it's useless to expect to go home.
My heart aches with grief;
My soul wanders around aimlessly.
Unable to make a living here, I'll try it in the East,
With a sudden change of luck, I may make it back to China.

In the early years, the population of Chinatown was predominately male. Many sojourners never made enough money to return to their families in China. Unable to bring wives and children to the United States because of immigration restrictions, these laborers often led lonely lives. They stayed in small boardinghouse rooms that were, according to one outsider:

Usually about ten by twelve feet in size, with a ceiling from ten to fifteen feet high. On two sides of the room (and sometimes on every side) bunks are placed one above the other like those arranged in state-rooms of steamboats. In a room where the ceiling is of ordinary height there will be three or four of these bunks on a single side. These are all occupied as beds for sleeping, from floor to ceiling. A small rental rate for each occupant yields a considerable sum for the room; it is, therefore, easily comprehended, how a Chinaman can afford to work for less wages than a white man.

Arnold Genthe, a San Francisco photographer, took many pictures in Chinatown in the early 20th century. He titled this one "The Street of the Gamblers." In a predominantly male society, gambling houses flourished, for playing such games as mah-jongg and fan-tan was a favorite way to pass the time.

This "bachelor society" survived for generations. Louis Chu, in his novel Eat A Bowl of Tea, *described the lonely existence of a bachelor named Lee Gong in the 1940s.*

Someone called "Uncle Gong" to him on the street but he did not hear him and continued walking toward his apartment. He climbed the stairs slowly, sighing heavily with each step.... Once inside he sank down on his single folding bed and buried his head in the pillow.... He was breathing heavily, like a physically wounded man. It was strange to find himself in his room at this time of day. He should be at Wah Gay's clubhouse playing mah-jong, which was what he had planned to do after his cup of coffee. He stared absently at the coal stove with its blackened chimney sticking into the wall.

He had lived in this room for more than twenty years. Twenty years is a long time. Maybe this room is unlucky.... Look what happened to his old roommate Lee Sam...still at the hospital somewhere on Long Island....

Sam was working in a laundry when one night two men came in to beat and rob him...after that he was never the same...he kept saying that someone was after him...finally they had to come and take him away...maybe this room brings bad luck.

Some of the "bachelors" had in fact left wives and children back home. The Chinese sojourners intended to stay in the United States only long enough to make money and then return. But many never did. One Chinese immigrant in Oregon wrote the following letter.

My Beloved Wife:

Yesterday I received another of your letters. I could not keep tears from running down my cheeks when thinking about the miserable and needy circumstances of our home, and thinking back to the time of our separation.

Because of our destitution I went out, trying to make a living. Who could know that the Fate is always opposite to man's design? Because I could get no gold, I am detained in this secluded corner of a strange land. Furthermore, my beauty, you are implicated in an endless misfortune. I wish this paper would console you a little. That is all that I can do for now.

Some members of the "bachelor society" stayed in fongs, *which were communal apartments, often occupied by people from the same family or district in China. The* fongs *were social centers that provided emotional and financial help to their members. Betty Lee Sung described how the members of a* fong *supported each other.*

I was brought up in Washington, D.C. We did not live in Chinatown, but my father went there religiously every Sunday afternoon. Sometimes the whole family went, and Father would drop us off to visit some family in the area, and he would go on to the *fong*, which was only a sort of dormitory setup with one room set aside as a sitting room. Father stayed

A Chinese American commented on the changes produced by female immigrants after World War II:

"Before the war, Chinatown was very quiet, almost a no-man's land because the Chinese could seldom bring their families over. Most men came by themselves. They lived together in tongs and started working each day as the sun rose and they stopped working when the sun set. You hardly ever found a man who had a day of leisure. But nowadays Chinatown is crowded with women all out shopping. It's at least 10 times more prosperous now."

The "bachelors" of Chinatown lived in communal apartments like this one, preserved in the New York Chinatown History Museum. Such simple accommodations allowed the bachelors to save their money to send back to their families in China.

Community organizations in Chinatown served the needs of those who had grown too old to work. This old man, photographed in 1936, seems resigned to the fact that he would never return to his homeland.

there all Sunday afternoon, sometimes sitting, sometimes just standing around greeting the others who arrived and "shooting the breeze." Many Chinese use the *fong* as their mailing address, and they would go in to read their mail from the home village. News about births, deaths, marriages, illnesses, political events, and even gossip was passed around, so that everyone was informed about happenings in their village of China.

The *fong* sponsored a *hui,* which in almost all respects served as a banking institution for the members.... Those who elected to join the *hui* agreed to pay into the pool ten dollars per share for one hundred weeks for a total of $1,000. Members who needed current funds would bid for the weekly pool by submitting a written bid before 3 o'clock that Sunday.... In essence this was a cooperative banking system in which the members were both borrowers and depositors.... Because my father was widely respected and trusted, he ran the *hui* for his village *fong* for many, many years. To the best of my knowledge, there was no formal organization or charter to this banking arrangement. Members who wanted to join merely signed up, and when enough members indicated an interest, the *hui* began....

For two hours each Sunday, Father served as a banker. On occasions, he served as one of the mediators when grievances and differences between members were brought before the *fong*. Typical cases were complaints against opening a laundry or restaurant too close to an existing one, wrangling over debts, problems with their base and origin in China but which could only be settled here because the family elders or heads were in this country, and once in a while, family quarrels.

Pardee Lowe's father owned a store like this one, photographed in San Francisco in 1908. Lowe said that men gathered there to "recall happier days at home when they crowded the village inns and ancestral temples." A woman stands behind the counter at left—a rare sight in those days, when merchants' wives usually stayed out of sight.

Letters from home frequently reminded the sojourners of their duties to their families back in China—as this selection, received by a laundryman named Ho-mou, indicates.

Ho-mou, Senior Brother-in-law, to you I write with respectful greetings:

You have been keeping yourself silent so long and have not been sending money home and Second Sister [Ho-mou's wife] has begun to worry. How and what you are doing now we know nothing about.

You have not written home a single letter for the last year. What has happened? My Brother must be industrious and save his money. It is a bad idea to spend money freely.... As soon as you receive this letter, please let us know how you are getting along.

Brother, Ming-leong
May 4, 1938

Ho-mou, Senior Brother-in-law, to you I write with respectful greeting:

Your letter arrived January 12 with a check for three hundred and twenty dollars Hong Kong currency. Of this amount you allotted fifty dollars to Mother Wo [Ho-mou's mother], fifty dollars to Older Brother Ho-yin [Ho-mou's brother], ten dollars to Cousin Wai-wah, ten dollars to Nephew Kwong-pui, and the rest to Second Sister. I have divided it as you wished and I take this opportunity to thank you for them.

You have said also, my Brother, that I should advise Second Sister to be careful with her expenditures. I did mention this to her, my Brother. The money you sent home last time was spent for the sake of your daughter. There was no extravagance. You can understand that, please, so that there is no cause to grumble.

My brother has been abroad many years and you know how Mother awaits your return. She wants you to be industrious and rich and return home in the near future.

Brother, Ming-leong
January 12, 1939

Three men in a home for elderly Chinese in Hawaii, around 1950. In China, such men would have been able to spend their last years, respected and admired, in the company of their families.

The smile on this man's face indicates that he maintained the same optimism that brought the sojourners to the United States in the first place. Bachelors gathered in the fongs for companionship and mutual assistance, and some did prosper here.

駐金山中華總會館圖記

The official seal of the Chinese Consolidated Benevolent Association.

STRENGTH THROUGH GROUPS

Chinese traditionally write their names with the family name first because they believe that is the most important thing about a person. From the beginning, a Chinese sojourner in the United States belonged to a group that he could turn to in case of need. Very soon, family, or clan, associations were formed to give aid to members in need, to settle disputes, and to send back to China the bones of members who died.

Wei Bat Liu, who arrived in San Francisco in the early years of the 20th century, remembered that his clan association provided living space for members.

In 1913, all the cousins from the Liu family in my village had one big room so all the members could fit in it, and we slept in that room, cooked in that room, one room. Anybody who had a job had to sleep outside the room, because he could afford to rent space and get a bed for himself. Anybody who couldn't find work slept in the beds in this room. At the end of the year, all the members would get together and figure out all the expenses. The ones that slept in the room most were willing to pay a little more. But even the ones who didn't sleep there were willing to pay something for the upkeep of the room.

Officials of the Chinese Consolidated Benevolent Association were usually wealthy merchants, like this group, photographed around 1890. The CCBA, which claimed to speak on behalf of all Chinese in America, was more commonly known as the Six Companies.

Twenty years ago, Victor and Brett Nee found these family associations were still active.

Almost every American born Chinese we spoke with had had some contact or experience with his clan. At a banquet, we talked to children who had fled the crowded dining room to toss paper cups around on the roof. Most of them didn't know what the words "family association" meant, but downstairs their mothers, many of them suburban housewives, said this was the one Chinatown event they brought their children to every year, "just to keep them in touch with the old traditions." A librarian, whose family's three generations were all present at the banquet, had insisted her daughter write an essay for the event: "How Youth Can Help Our Family Association." Another woman...said that there were only two points she absolutely required of her children: that they register their names with the family association and that they attend the yearly banquet. "It just makes me feel better to know that no matter what big city they're in, they can always go to their own family association if they need help," she said.

Ho Yang, who arrived from China as a boy in 1920, was an officer of his family association when he was interviewed in 1979.

Probably the most important thing the association does [today] is to run the Chinese schools. It's to teach the Chinese culture and literature and that sort of thing. The trouble is the majority of the children don't want to go. We got the school going up to the eighth grade, but most mainly go just through sixth grade. Even my own three kids, they don't like to go to the Chinese school. They say, "Oh, Pop, what's the use of learning Chinese? We're in this country now."

District associations served much the same purposes as clan associations, but their unifying principle was place of origin, not kinship. Most of the earliest immigrants came from 12 districts in Kwangtung Province. Around 1851, sojourners from 6 of these districts formed the Sze Yup Association (or Company.) In the same year, people from the 6 other districts formed the Sam Yup Association.

The district associations had a great deal of power. They made sure that any member boarding a ship for home had paid all his debts, and they collected exit taxes to support the activities of the associations. In 1855, a member of the Sze Yup Association described the aims of the organization:

The object is to improve the life of our members, and to instruct them in principles of benevolence. The buildings [purchased by the association] are somewhat like American churches. The company furnishes beds, fuel and water to guests who remain but for a short period; also lodging and medicines for the infirm, aged and sick. Means are bestowed upon the latter to enable them to return to China.

Despite their benevolent purposes, the district associations occasionally battled each other.

A famous "Chinese War" took place in Tuolumne County, California, in 1854. A quarrel arose when members of the Yun Wo Company trespassed on the mining claim of the Sam Yup Company. The Sam Yups posted an official challenge on red paper:

"There are a great many now existing in the world who ought to be exterminated. We, by this, give you a challenge, and inform you beforehand, that we are the strongest, and you are too weak to oppose us.... We are as durable as stone, but you are pliant as a sponge."

Such an insult demanded a response. Both sides kept local blacksmiths busy making 15-foot-long pikes, pronged spears, and swords. Word spread through the surrounding communities, and about 2,000 spectators waited at a large field outside Chinese Camp for the action to begin.

The spectacle was awesome. To the accompaniment of gongs and drums, about 150 Yun Wo members and 400 Sam Yups charged toward each other. Possibly because they carried shields woven from straw, casualties were surprisingly light. Two Sam Yups were wounded, and two Yun Wos killed. The Yun Wos fell back in retreat and were judged the losers.

A 19th-century magazine artist depicted his version of a fight between two tongs.

Arnold Genthe photographed this tong official, who is closely followed by a bodyguard. The tongs took over such activities as gambling, drug dealing, and prostitution.

Other district associations were formed later, representing sojourners from other parts of China. Around 1880 leaders of six associations formed a larger, more powerful group, the Chung Wah Kung Saw, or Chinese Consolidated Benevolent Association (CCBA). In San Francisco it was commonly known as the Six Companies.

The CCBA grew into a national organization, with branches in every city that had a sizable Chinese population. It became the most powerful and respected organization of Chinese Americans. The CCBA led the fight against the passage of the Chinese Exclusion Act. Though it failed in that effort, the group retained its enormous power until well after World War II. Most other Americans regarded the CCBA as the "government" of Chinatown, which spoke for all Chinese Americans.

A third type of Chinese American organization was the tong. Unlike the family associations and the district associations, the tongs were secret organizations. Open to members of all clans and home districts, the tongs were often associated with criminal activities. They controlled illegal gambling, prostitution, and opium smoking within Chinatown. Until the 1930s, members of rival tongs sometimes battled in the streets with hatchets or knives. These hatchet men, also known as high-binders, were seldom prosecuted by the police, for they followed a strict code of silence about their activities.

In China, secret societies also operated outside the law—but supposedly with a patriotic purpose. Their members worked to overthrow the "foreign" Ching dynasty. One of the most powerful Chinese secret societies was called the Triad Society. It took part in the unsuccessful Taiping Rebellion and the Red Turban Rebellion. Some of the Chinese who fled to the United States in the 1850s had been members of the Triad Society. In 1853, they established an American branch known as

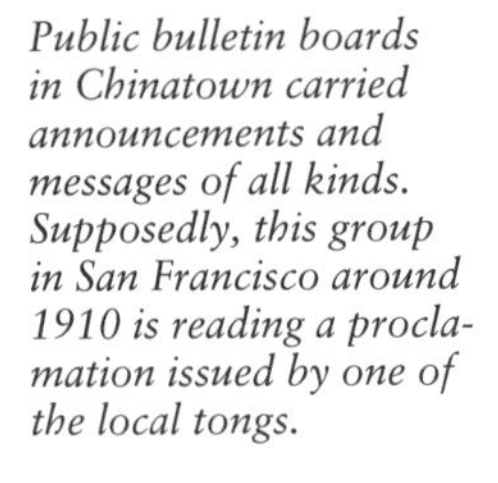

Public bulletin boards in Chinatown carried announcements and messages of all kinds. Supposedly, this group in San Francisco around 1910 is reading a proclamation issued by one of the local tongs.

the Chee Kung Tong. In the years that followed, many other tongs were organized in the Chinatowns of San Francisco, New York, and other American cities.

In 1972, Lew Wah Get, an 84-year-old man who was an officer of the Suey Sing Tong, described the reasons he had joined.

I first heard about the Suey Sing Tong in 1917. All the gambling houses were prospering around that time. I decided that if I went to work in one, I might make a better living than by working as a cook. I knew that in Stockton, especially, there were lots of gambling houses.... I learned that all these houses were owned by the Suey Sing Tong. So when I went to Stockton and became involved in one of the houses as a dealer, I also became a member of the tong.

If you wanted to join a tong, you had to have a friend who was already a member sponsor you. He had to swear to your good character, and even then the tong would investigate your name for one month before they let you in. This was the rule for everybody.... And once you were a member, you were on your honor to follow all the rules. If you did, then the tong would protect you. If anyone threatened you, or interfered with your business, the tong would help you out. Or if you couldn't find a job, the tong would send you someplace, or introduce you to someone who could give you work. This was why so many people wanted to join.

After you were in the tong for one year, you were eligible to become an officer. Only officers could participate in the business meetings, while the other members had no say. I became friendly with many different members and learned quite a bit about the affairs of the tong. So I was chosen to become a liaison officer, and held that post for over ten years. My job was to handle business between our tong and other tongs. If another tong wanted to have some transaction with us, I would discuss the matter with them, listen to their proposals, and then report back to our tong so a decision could be made. Fighting was a very frequent issue. If members of our tong had been threatened or their businesses tampered with, naturally we had to take steps to protect them. Or suppose another party owed us money and refused to pay, we might decide to bear a grudge and force retribution. Sometimes, when the other party kept refusing to pay, that's how trouble started.

A young man describes why he became a member of a Seattle tong.

When I left China my mother had begged me tearfully not to mix in tong affairs. They had a bad opinion of the tongs in my native country, not understanding their value. But I had been living among tongmen, they were my friends. I left school partly because of ill health and partly because I had suffered deeply from an unhappy love affair. In this state of mind, it did not take much urging from friends to make me join the Hip Sings.

The tongs had another function, as fraternal organizations of Chinese Americans. New York's Chinatown was host to a national convention of the Hip Sing Tong in the 1930s.

Henry and Ralph Jung, the sons of a Philadelphia family, in 1919. That year saw a rise in antiforeign feelings in the United States, and the flags may be intended to show the Jung family's patriotism.

FAMILY

The traditional Chinese family was based on the Confucian ideal of filial piety. The father commanded respect and obedience from his wife and children. Grandparents occupied a special place of honor. Tony Hom, who was born and raised in the United States, described how the relationship continued even after he became an adult.

It would just seem natural that you would take care of the elders once they reached that point in life. My grandparents were always close by. When we lived in an apartment, they were always a few floors up. Now, my parents have a two family house, and my grandparents live next door. So this was always pointed out to me. Maybe they are fearful that one day we would become too Americanized and not take care of them. You always hear them talking about how the "lo-fans" (referring to Caucasians) don't take care of the elders, while the Chinese community prides itself on how it takes care of its old family members. I was very happy and fortunate to be able to grow up with my grandparents. We shared many things like how they grew up. There were stories about how my grandfather hunted with his brother and caught fish [in China]. But they were so poor they didn't have oil to cook the fish in, things like that. They were in the position of authority and re-

A Hawaiian Chinese family around 1893. Hawaii had a higher percentage of Chinese families because the Exclusion Act did not apply there until 1902, four years after the United States annexed the islands.

spect. If I got a "no" answer from my parents, I could always fudge it and get a "yes" from my grandparents.

Right now, thank God, my parents are in a position where they can take care of themselves. I even talked this over with my wife before we were married. I know I would always take care of them regardless, and I was hoping that my wife would, too. And fortunately enough she stated that we would take care of my parents when the need arises.

A second-generation member of a Chinese family in Hawaii described her upbringing with affection.

Ours is a large family. Father had eight children; four boys from the first wife and four girls from the second wife. All of us children called the first wife "mother" regardless of whose children we really were. The second wife we call "Jah" which in Chinese means "second mother." This seemingly strange family situation can be explained by the age-old Oriental custom which allows a man to have more than one wife. Polygamy in the East is practiced by those who can afford it. A man takes a second wife for these various reasons: when his first wife fails to bear him children, when she fails to bear him a son, or when she is ill and incurable so that another woman is necessary to the household....

It is generally believed that in a typical Oriental family, the mother is secondary and unimportant, she being so submissive and meek. But this is not the case in our family. Mother was always the all-important mate of her husband. She was decidedly the ruling agent in the family, for father was too busy attending to his business (he was a merchant) and had little time to devote to the family.

Based as it is on a strict family system, an Oriental household is, economically speaking, carried on under communal principles. Our home is no exception. Our family is closely integrated, and we all work for the welfare of the home with that oft-quoted adage, "one for all and all for one," as the central motivating force.

Father was the economic head and all the things regarding financial matters were carried on under his direction. Mother and Jah took care of the details of running the house and bringing up the children, but it was father to whom they were ultimately responsible.

All earnings by the members of the family were turned over to father. My brothers' pay checks were given directly to father. Whenever anyone needed money, father was always willing to meet the demands provided they were not too extravagant.

From the time that we could help around in the house we were taught to be useful. We girls were taught to clean the house, help with the cooking, wash the dishes and in general to do all the household duties that a good daughter should know. Our brothers, on the other hand, were required to keep our yard well trimmed and well groomed. In addition they were

As Others Saw Them

A newspaper reporter described a Chinese temple to Confucius in Salinas, California, in the early part of the 20th century:

"Here Confucius sits enthroned in regal splendor. He is surrounded by a large and beautiful frame of wood, curiously carved.... Two brazen lions at the foot of the frame guard their kindly charge. In front of this god is a large stand holding five large lead vases. The front of the stand bears a wealth of carving, protected by a glass cover and strong wire screen. The carvings are illustrations of great men, like Washington, Foo explained, and also scenes in Chinese life, while the center is occupied by an illustration, showing Confucius dispensing justice. Here the faithful come and pray and make their offering of food and also money which is symbolized by the burning of a certain kind of paper. After each prayer the god is given a napkin in the shape of paper, as proper and necessary in connection with the food they gave him."

A family in New York's Chinatown early in the 20th century. The man's Western-style clothing indicates that he may have been a prosperous merchant.

Confucianism is not a religion in the Western sense. Joss houses included pictures of Confucius, his disciples, and other "sages," or wise teachers in Chinese history. Ceremonies included music and the presentation of a meal for the spirits of these teachers.

The home of a merchant in New York Chinatown in the early 20th century. Traditionally, a Chinese wife in such a family seldom ventured into the street except on holidays and other special occasions.

given the task of caring for the chickens that were kept in our back yard....

Serving tea is a very important function in the Chinese family. On our birthdays I would serve tea to all of the family. For example, if it were my birthday I would serve tea to everyone in the order of their importance in the family. Whoever receives tea gives me money wrapped in red paper. However, if it were my mother's birthday, she does not serve us, but instead, each of us serves her.

However peculiar may have been the household situation, I can say with sincerity that the happiness I found and the culture I received in my home are equal, if not superior, to the culture that could be got under any other family culture. I can say with a deep sense of pride and gratification that the teachings and training of my parents were of the highest order. Although educated in the American manner, it is my firm belief that my life will be guided by the truths taught me by my parents, for their teachings were sound.

I am an American-Oriental product, but it is my hope that my parents have not taught in vain and that when the sweet-scented incense burns before the family shrine it will bear to my forebears the message that I am fulfilling my task of carrying on my heritage of the East with honor and dignity.

In China, religion is closely entwined with family life. The idea of religion is different from what it is in Western countries. The word *chiao* means both "religion" and "education." There

is no feeling that one "truth," or doctrine, is superior to others. Chinese follow not only "the three great truths" (Confucianism, Taoism, and Buddhism) but also pay homage to numerous minor gods who bring good luck or ward off illness. Moreover, each home has shrines that honor the family ancestors. People went to temples—of any of these faiths—more to meditate rather than to worship.

The Hawaiian-born Chinese American quoted above recalled the practice of religion in her family.

Religion played an important part in the rituals of our family life. We had a separate room for the gods and goddesses with miniature shrines built for each. Every afternoon at three, the head of the household burned incense before the shrines. On the birthdays of certain goddesses we fasted and pledged ourselves to certain beliefs, e.g., if it were the birthday of the goddess of mercy, Kwan Yin, we prayed to her to help us to be merciful and to teach us to carry out her ideals in our world. From our earliest days we were taught the rituals and teachings of the gods by our parents, who always prayed and asked that we grow up imbued with the virtues of the deities....

Jah [her "second mother"] always celebrates the death of Father and Mother because she believes that their death on this earth signifies their birth in the next world. We always have a feast and shoot firecrackers to frighten the evil spirits away. However, when we children burn firecrackers, we think only of the fun of it and forget about the spirits.

As the second- and third-generation children of Chinese immigrants grew up, they became "Americanized." They acquired new attitudes that sometimes conflicted with the traditions of Chinese family life. Rose Hum Lee, a Chinese American sociologist, described the strains that arose between father and son in the Americanization process.

Eddie Wang's father could not make up his mind whether to be irate or amused.

"Look," he fumed as he waved a note under my nose. The note was found propped up against the telephone in the hallway when we entered the apartment. "Read this, will you." I took the note and read:

Dear Pop,

Gotta run back to school for the game. Need my black shoes for the dance tonight. Be a sport and polish them for me. Polish in upper drawer. Thanks.

Eddie

"Why the very nerve of that boy! Asking me—his father—to shine his shoes for him! Why, it's utterly disrespectful! When I was a boy, I spoke to my father only when spoken to. I stood in his presence. I took him his pipe and his tea. I dared not dis-

When my mother needed to go from one block to another in Chinatown, even from Mott to Pell Street, she would call for a coach. Women were ashamed to be seen on the street in those days.

—Mrs. Mok on her mother in the 1910s

Both Kong Tai Heong and her husband, Li Khai Fai, were doctors, educated in a British medical school in Hong Kong. In 1896, they immigrated to Hawaii, where they worked to improve health conditions among native Hawaiian, Chinese, and Portuguese laborers.

"I was very lonely. I sewed, read newspapers. *Gung Bao* and *Seung Bao* were the New York papers back then. Women were not allowed to help out in the store. They said that if women went down to the store, business would be driven away and men would stop shopping there. For a few tens of years I didn't go down to the store."

—Bing So Chin on being a merchant's wife in the 1920s

obey him. Nowadays, children have no respect for their parents," Mr. Wang wailed. "They do not pay any attention to the proprieties we observed toward our elders."

Yet, when Eddie came home, instead of sternly rebuking him or lecturing him, Mr. Wang treated the incident as a joke. To top it off, the shoes were shined and ready for Eddie in time for him to dash off to the dance.

That evening after dinner, I teased Mr. Wang about the shoe-shine incident. "It's your own fault," I chided. "You didn't reprimand Eddie for his disrespect. You even shined his shoes for him. I noticed that he didn't think it was anything out of the ordinary. To him, it was a perfectly natural request and you complied with it."

"What could I do?" lamented Mr. Wang, "Eddie was

A sizable number of Chinese Americans converted to Christianity, sometimes before their arrival in the United States. These girls are attending Sunday school at New York's Grace Faith Church in 1937.

brought up in this country. The customs here are different. With us, a father was a god. The cardinal virtue was filial piety. In the United States, the emphasis is on an easy camaraderie between father and son. My problem is I was brought up under a totally different culture. Sometimes Eddie's actions appall me, but we enjoy a warm relationship that I never experienced with my father."

Mrs. Teng, whose mother sold her to pay for her husband's funeral, found happiness in a marriage in Hawaii.

I believe the turning point of my life came when I was eighteen. One morning I overheard my master scold my mistress for wanting to marry me off to a man not of my same group. He said that long ago my mother made him promise that I be married to someone of my own group—Pun Dee [Punti]. He said that it is only fair to present the recent case [suitor] to me. I hurried away from the door and waited to be called any minute.... My master who was always nice to me said...that merchant, a Mr. Teng, from Wailuku, Maui, is looking for a bride. He is well-to-do but is forty years old. You are only eighteen. I leave the matter up to you. If he told me that the man was sixty I would have gladly said "yes."...

As a fee for my master's successful matchmaking my future husband sent him one hundred fifty dollars, a roast pig, five hundred cakes, a half dozen bottles of wine, and a half dozen chickens. All day I was buying things to take up to my new home. A lady took me down to the boat and when I landed at Kahului I was met by my brother-in-law who took me home to my husband. I became Mrs. Teng. My husband was almost bald but he was very nice to me.

Right after my marriage I asked my husband to write back to my village in search of my mother.... He sent my mother fifty dollars along with that first letter. I was very happy that I cried when I received my mother's letter telling me that my brother is eleven and is watching cows. I wrote home and sent her money to send my brother to school. I only longed to see my mother again. I think I would fall in her arms and cry for days but I never had that chance. She died a year after my husband's death in 1921.

The young people of today are very much changed. I cannot understand my daughter-in-law who never trusts me with her son. I am his grandmother. She is so afraid that I might put germs on him. When I have a slight cold I cannot go near him. How can I put germs on him? If he is healthy he gets no germs. The small children in China don't have enough to eat and no clothing and yet they don't die. The children in Hawaii have all the good food and clothing so why should they get sick?

A rare outing for a San Francisco Chinatown family around 1910. Photographer Arnold Genthe found them waiting for a streetcar, probably headed for a city park along the waterfront, where Chinatown residents went to enjoy the scenery.

THE CHINESE LAUNDRY

Chinese American laundries popped up almost everywhere in the United States—even in this little town in Arizona, where Native Americans waited outside Sam Kee's establishment for their clothes to be washed.

Prejudice limited Chinese to "acceptable" occupations such as laundry work. In addition, opening a laundry did not require much capital, for all the work was done by hand. Few of the "Chinese laundryman's" customers gave any thought to the hardships he suffered. From the 1930s to the 1950s, Paul Siu, the son of a laundry worker, interviewed people who made their living in the business. He recorded a typical day in a laundry in a Chicago suburb, June 10, 1940. The four men who operated the laundry lived there as well.

Tong was first up this morning," said Wah. "It must have been half past five. I don't know why the devil he has to get up so early.... As soon as he got up, he made a noise so loud with washing basin that it woke me up....

At eight, Tong was going out to collect laundry. Hong and Wah were working inside. The steam boiler was working and the wash tub was moving. Ming was alone in the office where he sorted and marked the laundry customers had just brought in.

The house was too noisy when the washing machine was moving; people had to shout to each other to converse. As soon

Laundry work was never an occupation for men in China. But in the United States, bigotry against Chinese Americans eventually drove them out of most other kinds of businesses. In the 1950s, Paul Siu said that Chinese took laundry work as "the quickest way to make money."

as the washing machine stopped, you knew the first tub of laundry was done and it was ready to be rinsed, wrung, and dried in a steady process....

About 8:30 Tong came back with his load in a wooden trunk attached to the tricycle. As soon as Tong put down the bundles, he left again, saying he had to make another trip. Ming came to open the bundles and began his sorting and marking work again. Hong and Wah were still inside, each busily at work, rinsing and wringing, and soon they put all clean laundry in the drying room....

About 10:00 the wheel of the washing machine turned again. Customers came to leave their laundry or to call for clean laundry. Tong and Ming were the ones who took care of the customers; these two spoke English better....

From the washing machine to the wash tub and to the wringing machine and the drying room takes about one hour and a half. Soon after the washing was done Hong began to cook for their lunch. Hong said sometimes he had no time to cook. He could cook today because they started earlier and the work was not so heavy. In case it was too busy, they would make some coffee and get some cold meat and cakes for lunch and they probably would not eat again until late at night.

The afternoon work constituted three main activities: starching, damping, and ironing of flat-work.... The ironing was done by old-fashioned irons [that were heated] on a gas stove.... Hong had about finished his starching and began to set collars and cuffs on a machine which was called *dai-goi* (big machine). Since every stiff collar and cuff had to go through his hands, it would take him the whole afternoon and deep into the night.

They did not eat their supper until 11:30 at night.... After supper, they all sat out in the yard to cool off before they went to bed. But they did not sleep until 1:00 A.M. It was a heavy meal, a big bowl of soup and large dishes of meat and vegetables.

A newcomer found the work hard.

I have been here about two months. No, I don't like it here. In China, people were talking about going to the "Flowery Flag" [America], and I was dreaming, too, about coming over. Now I am here. What I see in this country is just like this: working day and night.

I had thought our people were doing a big business here. I thought my brother had a big store. But all of us Chinese are just laborers. I have to work so hard in order to earn a small sum of money. If I had only realized the hard work I must do, I would rather stay home. I would rather stay in the village, feeling content to be a farmer. I don't know how to iron well enough yet. I am not allowed to iron shirts, as Shu-lung is afraid that I will spoil the shirt. If a shirt is spoiled, he has to pay for it. I can help to iron rough things such as underwear, towels, and shorts. Some people can iron 18 shirts an hour, but

As Others Saw Them

The New York Illustrated News *described a California Chinese laundry in 1853:*

"What a truly industrious people they are! At work, cheerfully, and briskly at ten o'clock at night. Huge piles of linens and underclothing, disposed in baskets about the room, near the different ironers. Those at work dampening and ironing—peculiar processes both. A bowl of water is standing by the ironer's side, as in ordinary laundries, but used very differently. Instead of dipping the fingers in the water and then snapping them over dry clothes, the operator puts his head in the bowl, fills his mouth with water, and then blows so the water comes from his mouth in a mist, resembling the emission of steam from an escape pipe, at the same time so directing his head that the mist is scattered all over the place he is about to iron. He then seizes his flat iron. It is a vessel resembling a small deep metalic washbasin having a highly polished flat bottom and a fire kept burning continually in it. Thus, they keep the iron hot without running to the fire every five minutes and spitting on the iron to ascertain whether it is still hot."

A Chinese laundry played a key role in capturing the notorious bandit known as Black Bart. Between 1875 and 1883, the banking firm of Wells, Fargo & Co. was plagued by this cunning highwayman. Wearing a black mask, Bart would suddenly appear on lonely stagecoach routes. Aiming his rifle at the stagecoach driver, Bart ordered him to throw down the strongbox that contained gold and silver. Bart seemed impossible to catch, and even left behind poems (signed "the po 8") that taunted the Wells, Fargo detectives.

On his last robbery, however, Bart dropped a handkerchief that held a Chinese laundry mark. Wells, Fargo's security chief traced the mark to a laundry in San Francisco. He learned that the owner of the handkerchief was a respected mining engineer named Charles Boles. Boles was convicted of the robberies.

I don't think I can do even four. The iron is too hot. You might burn the shirt, if you don't know how to handle it....

Now I am here. My brother has spent several thousands of dollars for my trip. I have to work for years in order to help him pay back the debt. I wonder when I can pay the debt.

The great hope that drew Chinese to the United States often turned to despair. Paul Siu tells of one man who promised his wife he would be home in three years, but stayed abroad for thirty! In the words of a man who had been hard at work for several years:

Being a laundryman is no life at all. I work fourteen hours a day and I have to send home almost all my wages. You see, I have a big family at home. My mother is still living and I have an unmarried sister who is going to school. My own children, five of them, are all in school too.... I figure I send home about fifteen hundred dollars a year, at least, sometimes more....

People think I am a happy person. I am not. I worry very much. First, I don't like this kind of life. It is not a human life. To be a laundryman is to be just a slave. I work because I have to. If I ever stop work, those at home must stop eating.

If the laundryman had a family, they too shared in the long hours of work. Buck Wong, a laundry worker's son, recalled:

When it came to a decision of whether I should study for my test or help with the work, I always worked. I knew my father wanted me to get a good education. But at that moment he forgot all about it. I'd say that I wanted to go out to the library. He would say, "Why do you have to go so much?" He would never let me out. I really hated the place [the laundry].

Even today, the laundry business provides job opportunities for newly arrived Chinese immigrants. This man operates an ironing machine in New York in the mid-1980s.

RESTAURANTS

From the days of the California gold rush, Chinese restaurants have been popular with both Chinese and non-Chinese. A miner in 1852 described one of the earliest restaurants.

They serve everything promptly, cleanly, hot and well-cooked;...their own peculiar soups, curries, and ragouts which cannot be imitated elsewhere; and such are their quickness and civil attention, that they anticipate your wants and of course secure your patronage."

Though opening a restaurant required a greater financial investment than a laundry business, Chinese Americans found it a profitable way of making a living. Johnny Kan moved to San Francisco with his family in the early 1920s. After graduating from grammar school, he began working for a Chinese grocer. Later, he started a bakery and soda fountain, but his partners would not listen to his ideas. Finally, he hit on the idea that would make him a success—a really good Chinese restaurant.

I thought back on the success of the Chinese restauranteurs in New York and Chicago in the roaring twenties. They were operating huge-size places along the Eastern seaboard, and they were packing in four to five hundred customers at one seating on Saturday nights.... You know, Chinese from San Francisco would willingly go to New York or Chicago just to work as waiters in these places. The crowds were

In the 1920s, the Port Arthur restaurant in New York Chinatown attracted diners from all over the city. It was named for a city in northern China where Japanese troops defeated a Russian force in 1904. All Asians felt a sense of pride at the first defeat of a European country by an Asian one.

Today, many Chinese American restaurants are family-owned businesses. Mrs. Wu (center) and her husband operate Cousins, a popular restaurant on the West Side of Manhattan.

The earliest kind of Chinese American cooking was called Cantonese, after the port city in Kwangtung Province where most of the earliest immigrants came from. In recent years, newer immigrants have introduced other cooking styles, such as Szechwan, Hunan, and Fukien.

For a long time, the best-known dish in Chinese restaurants was chop suey. Almost certainly, chop suey was created by a Chinese American because it is not found in China. Writer Alexander McLeod gives one version of the origin of chop suey:

"One night a crowd of hungry miners went to a Chinese restaurant on Dupont Street [in San Francisco] at a late hour, and the proprietor was ready to close. This the miners refused to permit, and demanded food. Their attitude disgusted the restaurant owner, who...dumped together all the scraps left over from serving his regular customers that day, in the same manner as when serving beggars in Canton. He put a dash of Chinese sauce on top of the scraps and served it to his unwelcome guests. They did not know what he meant when he told them that they were eating chop suey, or 'beggar's hash.' At any rate, they liked it so well they came back for more, and in that chance way the great chop suey industry was established. Many more Chinese fortunes have been made from it than were ever made from their gold mining in California.... Chop suey restaurants are to be found in the big cities all over the world, except in China."

so large and the tips so good, they wouldn't even ask for salaries!...

[F]inally I realized that the reason there were no first-class restaurants in [San Francisco's] Chinatown was because no one ever bothered to study, and to teach their employees, how to run a really fine place. And nobody had ever tried to educate Caucasians to an appreciation of Chinese food. There were over fifty restaurants in Chinatown—papa-mama, medium-sized, juke and soup joints, tenderloin joints, and others—where the waiters just slammed the dishes on the table and cared less about the customer or what he wanted to eat. So we decided to launch the first efficiently operated and most elaborate Chinese restaurant since the collapse of the old Mandarin. Our concept was to have a Ming or Tang dynasty theme for decor, a fine crew of master chefs, and a well-organized dining room crew.... And we topped it all off with a glass-enclosed kitchen. This would serve many purposes. The customers could actually see Chinese food being prepared, and it would encourage everybody to keep the kitchen clean. Also, in those hectic pioneering times, in a party of six or eight persons, there would always be one who disliked Chinese food....

Well, with our glass-enclosed kitchen, we could say to the difficult guest, "Will you please do us a favor? Come and watch us prepare fresh food and see our *woks*" [the oval-bottomed Chinese frying pans]. It usually took me about fifteen minutes to educate a disliker when I could have him smell the aroma of fresh, barbecued pork coming out of the oven, or sizzling filet of chicken breasts in a *wok* with beautiful, fresh vegetables around, snow peas or chard toss-cooked for just a few seconds. Usually a man like this will end up being a real enthusiast. Why? Because he never knew how real Chinese food was prepared. Now he was no longer a chop suey believer.

Tommy Woo, who came to the United States with his family after the exclusion period recalled:

My parents have a restaurant business in San Francisco. They are very hard working people. They start at eleven in the morning and work twelve hours a day. They've had the same restaurant for over ten years now, doing the same things over and over, serving the same people over and over, again and again. My parents didn't really want to open a restaurant, but, to them, they had no choice because they were immigrants and couldn't speak a word of English. They had other jobs, of course, such as janitorial work washing the floor of big restaurants at Fisherman's Wharf, earning only fifty to seventy-five cents an hour. During that time, my sister couldn't go to school because she had to stay home and take care of me all day. So my parents opened a restaurant. The years have gone by; and, to me, my parents seem to be satisfied with their jobs. To enjoy is different than to be satisfied.

Dong Kingman

Dong Kwong came to America in 1900 and found work as a laundryman. He was not typical, for after 16 years in the United States, he earned enough money to return to China with his wife and five-year-old son, Moy Shu. Dong Kwong opened a clothing store in Hong Kong, and Moy Shu decided to help out by making chalk drawings on the sidewalk to attract customers.

Moy Shu's mother, Lew Shee, recognized that her son had talent. She sent him to a grammar school whose headmaster was an accomplished artist. The headmaster taught Moy Shu that there were two kinds of art in the world—the Chinese style, *hsieh-yi* (drawing an emotion) and the Western style, *hsieh-cheng* (drawing reality). He gave Moy Shu lessons in both.

When Dong Moy Shu was 18, he went back to Oakland, California, showing his birth certificate to prove he was a U.S. citizen. Around this time, he took the name Dong Kingman. He worked in a jeans factory and saved his money. Someone sold him a restaurant for $75, but it failed, according to Kingman, because he spent his time painting instead of cooking. He then moved to San Franciso and worked as a houseboy. In 1933, he gave the first exhibition of his work at a student show at the Art Center in San Francisco. Critics raved about his lovely watercolor scenes, and Kingman found that people wanted to buy them. His career was launched.

Dong Kingman (left) and James Wong Howe (right) during the making of Howe's movie of Kingman creating a painting.

Kingman has always painted with watercolors because his earliest training had been with the brush and inkstone that Chinese traditionally use to write with. His favorite subject, nature, was the traditional subject of Chinese painters for centuries. Kingman soon became recognized as a master of the form. In 1940, the Metropolitan Museum of Art in New York purchased one of his paintings—a recognition that he was among the best of contemporary American artists.

Soon afterward Kingman was awarded a Guggenheim Fellowship, which he used to travel through the United States studying the endless variety of landscapes in the country of his birth. Kingman did not confine himself to the beauty of nature; he also painted his impressions of urban landmarks, such as the Brooklyn Bridge and Wall Street. In 1954, his friend James Wong Howe, an Academy Award-winning cinematographer, made a short motion picture of Kingman creating a painting of a street in New York's Chinatown. The film was shown around the world as a teaching tool for young artists.

Though Kingman was much in demand for magazine covers and commercial art, he set aside time to teach his skills to others. He has lectured and given courses in colleges on both coasts. His paintings, which have won many prizes, today hang in museums all over the United States. Kingman has fulfilled his father's dream of success while enriching this country with his genius.

James Wong Howe

James Wong Howe was born in China around 1899. His father brought his family to the United States and opened a restaurant in Pasco, Washington. The only Chinese child in the local school, James was teased by the older boys, and he learned to use his fists to defend himself. After his father's death, James drifted into professional boxing. But he saw that a boxing career was a dead end.

One day in 1916, Howe wandered through Los Angeles, looking for work. He turned a corner and spotted a movie company shooting a scene. Curious about the brand-new movie business, he stopped and saw a friend working on the set. The friend told Howe it was a great job: "The pay is terrific. I only work twelve hours a day and I get ten bucks a week." Howe hung around the studios and got lucky when he was hired by Cecil B. deMille, a director famous for his spectacular costume dramas.

Working for deMille, Howe found his true vocation. Eating from snack trays set out for the cast, he used his salary to buy a still camera. He experimented with lighting and lenses, watching countless movies to learn techniques. When he felt he was ready, he boldly asked Mary Miles Minter, one of the superstars of the day, to let him take some publicity photos. Minter was astounded at the results, and from then on, she insisted that he work on all her movies.

In 1927, a new development changed the movie industry—sound. "Talkies" ruined the careers of stars who looked beautiful in silent films but whose voices were jarring. Among them was Mary Miles Minter. Once more, Howe found himself out of a job. He gambled his life savings to produce his own movie, called *Transatlantic*. It was a smash hit, and from that time on, James Wong Howe's career moved steadily upward.

Though he endured racial prejudice and "a certain amount of insults," his talent soon made him one of the highest-paid camera operators in Hollywood.

He had a particular genius for photographing female stars. As he said, his task was "to make old stars young, plump stars thin, ordinary faces beautiful, and to convert terrible defects into gorgeous assets.... It's just a matter of trickery with lights, shadows, and the right accent of make-up." No longer known as "the Chinese cameraman," Howe acquired the reputation of being the best in the business. At one point, a studio had to hire bodyguards for him because one famous female superstar was rumored to have hired thugs to kidnap Howe to work on her next picture.

In his long career, Howe received 16 Academy Award nominations, winning the coveted Oscar twice—for *The Rose Tattoo* in 1956 and *Hud* in 1963. He died in 1976, but film students still watch his work to learn how a motion picture should be photographed.

美洲日報
The CHINESE JOURNAL

Chinese American newspapers like this one often supported the efforts of Sun Yat-sen, who is pictured above. Contributions of money and supplies collected in Chinese American communities helped to overthrow the Manchu dynasty in 1911.

TIES TO THE HOMELAND

Chinese Americans have kept in close touch with events in their homeland. Chinatown newspapers, letters from family members, and news brought by merchants who traveled back and forth from the United States to China were all sources of information. The Chinese Americans shared the goal of freeing their country from the domination of the foreign Manchu dynasty. One of the leaders of the rebel movements that sprang up was Sun Yat-sen, who had immigrated to Hawaii when he was 13. John Jeong remembered the day when this famous visitor came to his family's store in San Francisco's Chinatown.

One day in 1909 our friend Mr. Wu brought Dr. Sun Yat-sen into the store. Of course I recognized him because I had seen his face in pictures. I knew he was fighting against the Manchus and he had already lost a few battles in Kwangtung Province.

Yes, I admired Dr. Sun even before I met him. There were so many people in Chinatown who agreed with him! At that time, you see, we Chinese seemed to be without a country. We were just the servants of the Manchus, just like slaves, doing what we were told. If we didn't obey, they cut off our heads. But Dr. Sun spoke to us about China and told us not to be afraid. He said we could defeat the Manchus even without any arms. If we were patient and strong, he said, his organization could take positions in the government quietly, and seize their arms. Then Chinese people could become free again.

I went to some speeches and listened to all of Dr. Sun's ideas. Some friends and I decided to join his party, so we invited him to talk with us and have dinner with us at the Kam Wah Restaurant. When dinner was over we went up to his room and he gave us each a paper to sign so we could become members of the party. I was just twenty-one years old then, you know!...

Then after we signed they told us about the party [the Kuomintang party that Dr. Sun had formed]. Did you ever hear about the signs we had to use? We learned that when you meet a stranger, you should check to see if he is a member of the party or not. First shake his hand and scratch the inside of his palm a little. Second, if he is a member, he will ask you, "What's the matter today?" And you say, "The world's affairs." Third, he asks you, "What kind of person are you?" And you answer, "I am a citizen of the Republic of China." Just as I was joining the party, the newspaper the *Young China Daily* was started, and I became very busy working there. I used to do setting and printing and sometimes even on Sunday I worked until one o'clock at night.

In March 1910, there was a big uprising in Canton. It failed and many of our members were killed. In July and August 1910, I decided to go back to help the cause. I was ready to do anything they asked. If I had been told to fight, I was ready to die for our revolution. But I didn't learn how to fight when I was in America. Our China was very poor then and Dr. Sun had told us that our main job in America was to raise funds. So...when I went back to China, the best thing I could do was work on a newspaper again.... In 1911 I came back to the States.

Sun's rebellion succeeded in overthrowing the Manchu dynasty in 1911. However, no strong government appeared to take its place. After Sun's death, his successor, Chiang Kai-shek, attacked the communists who had been among Sun's supporters. China's disunity and weakness made it vulnerable; its stronger neighbor Japan started to carve out Chinese territory for itself. During the 1920s and 1930s Chinese Americans staged demonstrations asking other Americans to stop trading with Japan.

Japan launched a full-scale invasion of China in 1937. Letters poured out of China describing the hardships. One read:

Uncle, Venerable One, I write to you with respectful greeting:

Now Canton is captured by the Japs, our commodities here cannot be shipped to the village. For this reason, the prices of foodstuff in the village are high, very high. One bag of rice costs from eight to ten dollars. How can the poor families back home manage to live!

However, everybody at home is well. I hope you are well, too, in America.

Nephew

Members of the Chinese Hand Laundry Alliance, the laundry workers' organization, contributed the funds to buy this ambulance for a Chinese hospital during World War II.

In another demonstration of support for their homeland, Chinese Americans carried this huge Nationalist Chinese flag through the streets of New York Chinatown.

To men of my generation, World War II was the most important historic event of our times. For the first time we felt we could make it in American society.

—Charlie Leong of San Francisco's Chinatown

Rose Ong, a seamstress in San Francisco, sent six sons to serve in the U.S. armed forces during World War II.

In 1941, the United States entered the war against Japan and its allies, Germany and Italy. Chinese Americans enthusiastically gave their support to the war effort. Second-generation Chinese Americans—who had been born here and hence were citizens—signed up to serve in the armed forces of the United States. Chinese American women, who had long shunned public activities, took part in war bond drives, volunteered as nurses, and took jobs in factories that produced goods for the war. The Chinese Americans of Portland, Oregon, contributed money to pay for three airplanes to fight the Japanese; the planes were named For the People, By the People, *and* Of the People.

The San Francisco Chinatown community issued "A Memo to Mr. Hitler, Hirohito, and Company":

Have you heard the bad news? America is out to get you. America has a grim, but enthusiastic bombing party started, and you're the target in the parlor game.

San Francisco Chinatown, U.S.A., is joining the party. Chinatown will have fun blasting you to hell. Chinatown is proud to be a part of Freedom's legion....

Chinatown's part of the party will cost $500,000. Admission price to the fun is purchase of a U.S. War Bond. We're all going to buy a War Bond for Victory.

Chinese American Boy Scout troops joined in the wartime drives to collect tin cans and other materials that could be used for making weapons.

A Chinese American teenager, Florence Gee, wrote an essay in the Chinese Press, *a newspaper, in 1942 that declared:*

I am an American.... The war has hit home. I have an uncle in the army and one in the shipyard. My sisters are members of the civilian defense. My mother is taking first aid. I belong to a club where I learn better citizenship.

The war was a turning point for Chinese Americans. Because the United States and China were allies in World War II, the Chinese Exclusion Act was repealed in 1943. A new age was beginning for Chinese Americans.

A "lucky" Chinese dragon leads this Chinatown parade during World War II.

This grandfather and grandson share a quiet moment.

CHAPTER SIX

PART OF AMERICA

After World War II, the "bachelor society" of Chinatown changed forever. In a complete reversal of earlier patterns, nearly 90 percent of new Chinese immigrants were female. Some were wives who had waited patiently for years until their husbands in the United States—now permitted to apply for citizenship—could bring them into the country. Others were young women who married Chinese American soldiers serving overseas. Families became more common in Chinatown.

As the younger generation of Chinese Americans entered the mainstream of American life, they developed new attitudes that brought them into conflict with their elders. Jade Snow Wong's autobiography, *Fifth Chinese Daughter* (1950), describes her struggle to attain independence from her parents and gain her father's permission to pursue a college education.

Similarly, C. Y. Lee's novel, *Flower Drum Song* (1957), describes the conflict between Chinese and American cultures. When the book was made into a movie and stage musical (by Rodgers and Hammerstein), it was transformed into a lighthearted, sugary version of Chinatown life. Some Chinese Americans found the movie and play as much a stereotype as Fu Manchu had been, but at least it was a favorable one. Arabella Hong, a graduate of the renowned Juilliard School of Music, made her Broadway debut in *Flower Drum Song*—probably the first Chinese American to take a leading role on the American stage.

Some prosperous Chinese Americans felt secure enough to move to the suburbs, as so many of their fellow Americans were doing in the postwar years. Monterey Park, near Los Angeles, has been called "America's first suburban Chinatown." Most of its residents are from Hong Kong and Taiwan and are wealthy professionals and businesspeople. During the early 1980s, the city elected its first Chinese mayor, Lilly Chen. Flushing, part of the borough of Queens in New York City, has seen a great influx of middle-class Chinese since the 1970s.

As racist attitudes softened, more Chinese American sons and daughters entered American universities and found success in occupations that had earlier been closed to Chinese Americans. Many new immigrants from Hong Kong and Taiwan already had a college education and found that their talents won them jobs in high-tech industries. Today, a greater proportion of Chinese Americans have college degrees than any other American ethnic group. But science is hardly the only field where Chinese Americans have made their mark. They have made important contributions in entertainment, communications, sports, literature, arts, architecture, and politics.

The ancient values set forth by the Chinese philosopher Confucius have been a guiding force behind the Chinese American success story. An Wang, the richest Chinese immigrant and founder of Wang Laboratories, attributed his business success to the "attitudes and values that I acquired in China." These values, which included respect for education and hard work, have helped others achieve the American dream of owning their own home and business.

The great spurt of new immigrants after 1965 created a much larger but divided Chinese community. The *lo wa kiu* ("old overseas Chinese") were in the majority in 1965. Today, the *san yi man* ("new immigrants") make up more than 60 percent of all Chinese Americans. Between 1980 and 1990, the Chinese American population more than doubled, largely because of immigration.

Few, if any, recent immigrants regard themselves as temporary residents or sojourners, as the old-timers did. Three-fourths of them become naturalized American citizens within eight years of their arrival. Although a great many of the recent immigrants have a college education and technical skills, others are forced into menial jobs and overcrowded living quarters in old and new Chinatowns. By some estimates, immigrants make up more than 80 percent of the population of New York's Chinatown today.

These people, like the 19th-century sojourners, take jobs that no one else wants, with low wages and desperately hard working conditions. Today, as before, garment workers are among the most exploited. Wing Ng, who came to the United States from China in 1965, said, "The conditions in the factories are terrible. Dirty air, long hours, from eight in the morning to eight at night, six days! They don't protest because they don't know how to talk back and they don't know the law."

Many take such jobs because they have arrived illegally, like the "paper sons," dreaming that once they have arrived they will find the riches of the "Golden Mountain." Others have risked their lives to escape communist regimes in Vietnam or China. They survive by working in sweatshops, selling wares on the street, or trying the old, familiar trades of laundryman and restaurant worker. Starting at the bottom of the ladder, they hope to follow in the footsteps of the Chinese Americans who have "made it."

Chinese students in a New York City high school perform a traditional dance as part of a school celebration of Asian cultures.

Many prosperous Chinese Americans retain sentimental ties to Chinatown, though they may live in middle-class suburbs. Using their political and economic clout, they have encouraged city governments to build better housing and provide social services for the newcomers. Politically aware Chinese American college students demonstrate to win rights for the less fortunate recent immigrants. Chinese Americans, however prosperous, are aware that discrimination against them is still alive.

Today's Chinese Americans have taken charge of defining their identity. What are Chinese Americans like? For a century, this question was usually answered by other Americans, sometimes with sympathy, more often with bigotry. In recent years, many Chinese American scholars have begun to research the history of their forebears. Rose Hum Lee led the way with her sociological studies of Chinese American life. Thomas W. Chinn, Him Mark Lai, Victor and Brett de Bary Nee, Ruthanne Lum McCunn, Betty Lee Sung, and Jack Chen are among those who have made significant contributions in recent years to the history of Chinese Americans. The Chinese Historical Society of America in San Francisco and the Chinatown History Project in New York are only two of the many groups that are seeking to preserve and rediscover Chinese American culture and history.

Chinese Americans in the arts have led the effort to replace the old stereotypes with the voices and faces of real individuals. Maxine Hong Kingston and Amy Tan, two of the best-known Chinese American novelists, have eloquently described the struggles of Chinese American families and explored the conflict between generations.

Frank Chin, a fifth-generation American playwright, has ferociously challenged the stereotypes that have plagued Chinese Americans. Chin coedited two influential collections of Asian American writing—*Aiiieeeee!* (1974) and *The Big Aiiieeeee!* (1991). The titles refer to the fact that American popular culture —movies, TV, comic books—"pictured the yellow man as something that when wounded, sad, or angry...whined, shouted, or screamed 'aiiieeeee!'" Chin and his coeditors condemned most earlier (and some contemporary) literature by Chinese Americans as "snow jobs pushing Asian-Americans as the miracle synthetic white people." In the two *Aiiieeeee* anthologies, the editors presented the work of writers who "are elegant or repulsive, angry and bitter, militantly anti-white or not, not out of any sense of perversity or revenge but of honesty.... We are showing off. If the reader is shocked, it is due to his own ignorance of Asian-America. We're not new here. Aiiieeeee!!"

Wayne Wang, a young moviemaker born in Hong Kong, presented an insider's look at an American Chinatown in his film *Chan Is Missing* (1981). Six years later, Wang produced *Dim Sum: A Little Bit of Heart*, a warm portrayal of the relationship between a tradition-minded Chinese American mother and one of her daughters. Wang's most recent film, *The Joy Luck Club* (1993), based on the novel by Amy Tan, won widespread acclaim from critics.

However far Chinese Americans have moved from old Chinatown, many still celebrate the traditional New Year's festival, which is as much a part of America as St. Patrick's Day. It has remained a link with tradition for many Chinese Americans. Today, they return to celebrate as long dragons or lions with fierce heads parade through the streets of modern Chinatown. Because each Chinese family has its own feast for relatives and friends, Chinese restaurants close for the holiday, cleaning and renovating. Other merchants give away candies and treats to children. All debts must be settled before the New Year begins.

Even in the 19th century, this was the one time of the year when other Americans came to Chinatown to watch the festivities—and it remains a festive attraction today. Though outsiders might not appreciate the deeper religious significance of the festival, no one can fail to enjoy the boisterous and happy public celebration of New Year, with its hope for better times.

Chinese New Year has become so much a part of American culture that in 1993 (4691 in the Chinese calendar), the U.S. Postal Service issued a postage stamp to mark the occasion. It pictured a rooster, which was the symbolic animal for that year—the 10th in the 12-year cycle of the Chinese calendar. The animal for each year—rat, ox, tiger, snake, horse, dragon, among others—has a particular significance that determines what kind of year it will be. By tradition, children born in that year take on the characteristics of its animal.

It has taken a long time for Chinese Americans to gain the acceptance symbolized by that postage stamp, although many say acceptance is not yet complete.

Top: War brides attend a dinner of welcome in 1946. Bottom: Chinese American women organized groups that took part in feminist and antiwar demonstrations.

Today's immigrants, some of whom have come to this country illegally, are often forced into jobs with substandard pay and working conditions. This woman's child stays with her while she toils in a garment sweatshop.

SWEATSHOPS AND HIGH TECH

More than 60 percent of the residents of Chinatowns in the 1990s are new immigrants. Refugees from the People's Republic of China and the countries of Southeast Asia often arrive without funds, eager to take any job they can find. Emily Young, the daughter of an immigrant who found a job as a garment worker, told her story.

My mom used to work at a sewing factory. Now her health is failing so she can't work at all. But she always says that, well, some of the bosses are good, but a lot of Chinese bosses...say, "Fine, fine. Why don't you take another batch and do it while you're home on your free time?" Which is supposed to be illegal, of course. At any rate, whenever you're out of favor in a garment factory, the boss will always tell you your stitching has to be done over. Which really hurts, because even if they say they're paying by the hour, most factories pay by the piece.

Fortune cookies, a Chinese American invention, also provide a way for new immigrants to earn a living. But the cheerful messages inside the cookies give no indication of the desperate lives of the workers who made them. A girl who came from Hong Kong in 1972 describes her mother's first employment in San Francisco.

After a few months, one of our relatives walked by a fortune cookie factory in Chinatown. They were looking for a Chinese worker for a full-time job. She told my mom right away and the next day, she was accepted. So from then on, my mom became a factory worker. In the beginning, she didn't have any thick gloves and her fingers were always burned by the hot cookies. The temperature in the factory is terribly hot, especially in the summer. It's stuffy and noisy inside. It takes a long time to practice the skill of folding and bending the cookies. During these times, the beginning workers don't get full pay. They have to wait 'til they can handle a machine all by themselves. Eight and a half hours was too much for my mom. Besides she didn't get used to conditions there. So later, I began to go help her and became a fortune cookie worker myself. In my opinion, I hate this job absolutely. It did nothing good to us except harm of my health and time. My mom gets sick very often. But she can't ask for a day off unless she is very, very sick. So sometimes she has to work even though she has a toothache. Few years ago, they all had a day off on Sundays. Now, they have to work seven days a week. After work, my mom has to sweep the floor without pay. We have to make our own gloves and mend them after we go home.

Chen-Ning Wang and Tsung-Dao Lee

Since World War II, many college-educated Chinese have immigrated to theUnited States. Two of them, Chen-Ning Wang and Tsung-Dao Lee, shared the Nobel Prize for physics in 1957. They were the first Chinese Americans to win a Nobel Prize. Wang was born in 1922 and Lee in 1926. After Japan invaded China in 1937, their families were among the millions of Chinese uprooted by the fighting. In 1945, the last year of the war, the two young men met for the first time while students at a Chinese university in Yunnan Province.

After the war, Wang and Lee both won scholarships to study physics at the University of Chicago. By the early 1950s, they were working at the Institute for Advanced Study in Princeton, New Jersey. The institute brought together some of the most brilliant scientific minds of the time.

Lee left the institute in 1953 to become a professor at Columbia University in New York City. He and Wang frequently met for dinner at a Chinese restaurant in New York. There, they developed the idea that would win them the Nobel Prize. In 1956, they startled the scientific world by announcing the results of their experiments. When the Nobel Prize Committee honored Wang and Lee in 1957, they were two of the youngest men ever to receive the coveted prize.

When Dr. Chen-Ning Wang accepted the Nobel Prize, he said, "I am proud of my Chinese heritage and background, as I am devoted to modern science—a part of human civilization of Western origin to which I have dedicated and shall continue to dedicate my work."

Two other Chinese Americans, C. C. Ting and Yuan T. Lee, have won Nobel Prizes in science since then.

An Wang

When An Wang arrived to study in the United States in 1945, he lived on $100 a month. By the time of his death in 1990, his fortune was measured in the hundreds of millions of dollars. "One thing I have discovered," he wrote, "is that attitudes and values that I acquired in China...have had a great bearing on the way I do business. These values have much in common with some of the virtues of Confucianism."

This remarkable man was born in Shanghai in 1920. A university student when Japan invaded China, Wang spent the war years building radios for the Chinese troops. In 1945, he came to the United States for further education. "I had heard that there was discrimination against Chinese in the United States," he recalled, but he was confident "that I could acquire whatever skills I needed to survive here."

Wang enrolled at Harvard University, where he saw the first computer built in the United States, the Mark I. Though it was 51 feet long and 8 feet high, Mark I was little more than a gigantic calculator. An Wang joined the group of scientists who worked to improve the unwieldy machine.

One of the scientists' most perplexing problems was how to create a memory in which the computer could store information. In 1948, "while I was walking through Harvard Yard," Wang recalled, the solution "came to me in a flash."

In 1951, he started his own company, Wang Laboratories. Its first products, desktop calculators, cost $6,500 each, but businesses eagerly snapped them up. Wang poured much of his profits into developing a word processor. Wang Labs was among the first companies to combine a computer, a video screen, and a keyboard into the machine that is now found in every school and office.

After his company became a billion-dollar business, An Wang shared his fortune. He donated millions to Harvard; to the Chinese Cultural Institute in Boston; to provide scholarships for students from China; and to build hospitals and arts centers. Wang wrote in his autobiography, "When we enter society at birth, we receive an inheritance from the people who lived before us.... I feel that all of us owe the world more than we received when we were born." No Chinese American has done more for his adopted country.

Michael Chang

Tennis fans watching the final match of the 1989 French Open felt certain that Michael Chang's amazing string of victories was about to come to an end. Earlier in the tournament, the slightly built, 17-year-old American had defeated Ivan Lendl, the number-one player in the world. But now, in the final match against the number-three player, Stefan Edberg, Chang seemed tired. After winning the first game, he lost the next two.

But Chang fought back. Time after time, he returned Edberg's powerful serve. The match dragged on for hours. Chang eked out a win in the fourth game to tie the score. Then, in the final game, he broke Edberg's serve and held his own for a victory.

Michael Chang was the youngest man ever to win the French Open and the first American to take home the silver victory cup in 34 years. Chang modestly credited his success to the support of his family. Even before the French Open, sportswriters noticed that his father, mother, and older brother were always at his matches, offering advice and pumping up his confidence. A sports magazine dubbed them "the Chang Gang."

Joe Chang, Michael's father, came to this country in 1966 from Taiwan. A research chemist, Joe Chang developed an interest in tennis, enjoying it so much that he moved his family to California so they could play year-round. His younger son Michael showed great promise, winning his first tournament when he was only seven.

Joe Chang took a scientific approach to training his son. He made graphs and flowcharts to mark Michael's progress. "Creating a tennis champion," he said, "is 90 percent information gathering and 10 percent creativity. The important thing is to have the right 90 percent information."

After Michael Chang's victory in Paris, he suffered a hip injury and had to follow a tough program of physical therapy to regain his strength. Since then, Chang has continued his career, and according to his father's charts, he still has not reached the peak of his potential strength and ability.

MAKING IT

The immigrant's long struggle for success and prosperity sometimes paid off. Arthur Wong obtained a job with a laundry when he arrived in the United States in 1930. He wanted to work as a waiter, but could not speak enough English. Some friends advised him to buy a Chinese-English dictionary. It was the beginning of his road to success.

Well, I took the advice of these young fellows and I got myself a dictionary. I think I carried that dictionary for about three to four years, in the back of my pocket, just like a pack of cigarettes.... I walk on the street and I see a word and I bring out my dictionary and I find out what that word means. So when I learn one word and two and three, in time I build my own sentences; and I learn my own language that way. If you accumulate words like the way you put money in the bank, two words in your notebook a day, 365 days a year, you will learn over seven hundred words in one year....

After a few years in New York...I went to work as a waiter, a part-time waiter. And I work seven days! I work five and a half days in the laundry and work the whole weekend in the restaurant. And then came the war, and defense work open up.... So I went to work for Curtiss-Wright, making airplanes. I started out as an assembler, as a riveter. By the time I left for the army, I was an assistant to the foreman.... I got a deferment from the armed services until the situation got critical, and then I got drafted by Uncle Sam—said, "Come here, son!" Well, actually, in those days you felt privileged to handle a gun to defend your country in World War II.

When I got back...I hear there's a laundry shop for sale, and through friendship and grapevine I find this little laundry in New Jersey. And we looked at it and we bought it and we started our really long, successful journey till today. We work at it for twenty-seven years, and I think we did pretty well.

Practically most [of] our dreams has been fulfilled, up to this point. I came here to do what I like to do and hope to do, and I've been fortunate. With a little luck, we have accomplished. I got a family and started out, big long struggle, and own my own home, raised the children, educated them. They are all coming along. First-born was a boy, now twenty-eight years old. He's already got two degrees in his pocket. He's now an American diplomat in the foreign service. My daughter is doing social work, as a hospital consultant for patients. And the youngest one, twenty-two years old, he just graduated from college and was awarded a full fellowship to study marine biology.

Now I'm retiring from my long struggle. Certainly I don't think there's any place in the world we could do what we did, with what I have. All I have is ten fingers. I have no money, no education. But I know I have one thing—an opportunity to prove what a man could do.

Wong Chun Yau, who fled the terrors of China's Cultural Revolution, describes how she started a new life.

My daughter lived in San Francisco, so that's where I went when I arrived in this country in 1979. I was sixty years old. There was a social service agency there that provided orientation and training for new immigrants. [The teacher] would send me to a job, where I made three dollars and twenty-five cents an hour for three hours. I said, "Wow, more than three dollars an hour." I had never made so much money in my life. I was ecstatic....

In San Francisco I was sent to a hotel to clean bathrooms and pick up cigarette butts. It was work, and I couldn't believe I had such an opportunity to make money.... I say, this country is great. There is no comparison between China and the United States.... Not only did I make more money here, I could also buy whatever I wanted. How great it is. The U.S. government is excessively wonderful....

My sons have all prospered here, especially my youngest. He owns over six buildings. So, what is there to worry about? Nothing. My eldest son has three buildings. He gave me five hundred dollars for Chinese New Year this year....

At this senior citizen center I go to, I don't have to cook, and there is a meal for me. When I am finished, someone even wipes the table.... I even got a picture from President Bush thanking me for voting for him. I don't know which party he belongs to. I just voted for the one who would be most helpful for old people.

Mrs. Kong was one of the many thousands who arrived after the 1965 immigration-law reform. She escaped from the People's Republic of China in 1972 and settled in New York's Chinatown, which is now the largest in the nation. Starting as a street peddler of fresh vegetables, Mrs. Kong saved enough to open a shop. Today, she and her husband own four shops, in which most of their employees are family members. However, as Mrs. Kong's daughter Tina related, everyone still works hard.

This family business. My mother run around Chinatown, she always run around. I'm helper here, twelve hours a day before I'm pregnant. My mother works seven A.M. to nine P.M. almost sixteen years. No day off in almost sixteen years. Only New Year's only. Cousins never have a day off—they don't want it, because they're new immigrants, and also the financials not that well, maybe they need to save money for their own house or a store. My grandmothers work seven days a week—almost ten hours a day. They make bean curd and cook lunch for workers. Anything—they can handle it. Ages seventy-six and sixty-two. They like it. They strong, they never sick. No sit. Only stand up, ten hours a day.

Connie Chung

In May 1993, Connie Chung reached the top of her profession as a television journalist. Only 46 years old, Chung was named cohost of the "CBS Evening News." This promotion capped a 22-year career that had already made her one of the best-known Chinese Americans. Millions of TV viewers know Chung's face from the many programs on which she has appeared. She has won three Emmys and a Peabody, the highest accolades of broadcast journalism.

Born in Washington, D.C., in 1946, Connie Chung was the 10th child of a family that had fled war-torn China during World War II. Her childhood in the suburbs of the nation's capital helped form her career. "You can't grow up [where] I did," Chung recalled, "without developing an interest for how this country works."

After graduating from the University of Maryland, she took a job with a D.C. television station, where she was assigned to cover murders and plane crashes. It was, she recalled, "quite a shock for a girl from a sheltered Chinese home. But I'd plow through and get there anyway."

In 1976 she moved to the CBS station in Los Angeles, where she got her first anchor post. During the next seven years she won several awards for her reporting and became the highest-paid local news broadcaster in the country. Even so, in 1984 she moved to the NBC network, which offered her the chance for nationwide exposure.

In 1987, she visited China for the first time, broadcasting live from Beijing, where she interviewed some of her own relatives, whom she had never met before. "It was the most rewarding experience I've ever had," Chung says, "Through their experience, they told the history of modern China—how the war affected this family.... I went to my grandparents' graves...and I cried a lot with my relatives.... My life has been much more defined by my roots since that experience."

In 1989 she returned to CBS to host "Saturday Night with Connie Chung." Though the program did not last long, CBS had bigger plans in mind for her. Her latest post as co-anchor of the "CBS Evening News" will provide new opportunitites for her talents. Those who know her best think that Chung's greatest achievements may still be ahead of her.

I. M. Pei

The Rock and Roll Hall of Fame in Cleveland. A glass pyramid in the courtyard of the 800-year-old Louvre museum in Paris. The John F. Kennedy Library in Boston. The National Gallery of Art in Washington, D.C. The Mile High Center in Denver, Colorado. The 72-story Bank of China in Hong Kong, the tallest building in Asia. What do all these buildings have in common?

These and countless other striking and unusual structures were all designed by I. M. Pei, one of the most famous American architects. Born in Canton on April 26, 1917, he was given the name Ieoh Ming, which means "to inscribe brightly." Hoping to become an architect, Ieoh went to the United States in 1935.

By the time he graduated from the Massachusetts Institute of Technology, Japan had invaded China, and Pei decided to stay in the United States. William Zeckendorf, a New York real estate tycoon, gave Pei his first chance to make a mark on the urban landscape. Working with him, Pei designed housing projects and shopping centers in New York, Denver, Montreal, Honolulu, Washington, and Chicago. Zeckendorf taught Pei the importance of city planning—making sure that the rapid growth of urban areas would not destroy the quality of life.

Pei opened his own architectural firm in 1955. Though Pei continued to develop offices and apartments, he won commissions for other buildings that allowed him to give full rein to his creative ideas. One of his guiding principles was that a building must fit into its environment. A science center he designed in Boulder, Colorado, was made of red concrete towers so that it would blend with the sandstone buttes in the background.

In the 1980s, he was particularly proud to receive the commission for the Bank of China skyscraper in Hong Kong, for his father had once been manager of a branch of the bank. Today, two of Pei's own sons have joined his architectural firm, giving the promise that the name Pei will continue to be associated with great buildings into the next century.

BECOMING AMERICANS

The racial barriers that blocked Chinese from becoming American citizens have fallen, and each year thousands of immigrants go through the naturalization process. Steven C. Lo, in his book The Incorporation of Eric Chung, *described the immigration test of Eric Chung, a fictional Chinese immigrant in the 1980s. In some ways, the process seems as confusing as the grueling examinations that the "paper sons" faced. But it proves to be considerably easier.*

There was an exam—the Immigration called it "Naturalization Interview"—which was designed to show I knew enough about making a living here and would not be a "burden" of some kind.... The immigration officer, Mr. Shugart...began his test with inquiries about my past.... Then he wanted me to assure him, "in my own words," that I had always paid the taxes, obeyed the laws, stayed away from the Communist Party, and so on. After that he looked up very officially at the space above my head and rapid-fired:

"What is the function of the judicial branch of the American government?"

"What is the purpose of a jury?"

"When a person is accused of a crime, what is he entitled to?"

A mural in New York City shows two young Chinese Americans contemplating the present as they move toward the future.

These questions were all from the "exercises" part of the book [that Eric had studied], so I was just as quick with the answers.... After about ten minutes of what you'd call "intense questioning," he became more relaxed and sank low into his swivel chair. The blue-paneled cubicle was small but comfortable....

There was a heavy smell, a mixture of coffee, cigarettes, and old papers. He stopped to check his folder.

"Good...fine," he said, "now we only need to prove you have enough knowledge about English."

"Yes, sir," I said, waiting. He stayed silent. "Maybe...maybe you want me to discuss some works in American literature?" I asked.... There was still no response from him. He checked into his folder again. "I can discuss a couple of poems, maybe," I said, and seeing that he was still unimpressed, I made a more specific, and daring offer: "Poems of Wallace Stevens. I can, maybe, talk about the meanings of several of his poems?"...

"Well..." the man said.

"Yes?"

"Can you write something for me?" he asked, and handed me the sheet in his folder. It was a page of all the questions he'd asked me. There were checks, markings, and his written comments all over the place. At the bottom of the page was a blank line. He pointed at the line and said, "Can you write 'I love America' for me?"

"I love America?" I asked.

"I love America."

I did what he said, taking the time to make sure of my neat penmanship. He read my writing and put the paper back in the folder.

"Good," he said, smiling for the first time. "We now have everything we need." He nodded. That's when he stood up and congratulated me.

The names and physical appearance of Chinese Americans still cause people to mistake them for "foreigners." Kie Ho, a businessman who lives in Laguna Hills, California, wrote in 1982:

At a recent seminar that my company sponsored, where many of the participants came from our overseas offices, a gentleman from the Netherlands looked at my name tag and said, "I see that you are from our division in California, but your name does not sound American." I told him that mine is indeed a Chinese name; however, I am an American citizen.

I should have told him that my name is as American as Lucille LeSueur or Margarita Carmen Cansino before they became Joan Crawford and Rita Hayworth. My name does sound as foreign as the name of the Japanese slugger Sadaharu Oh, but does it not also sound as American as Joe DiMaggio?

I have already simplified my name for Yankee ears. I was born Kie Liang Ho, which means Ho the First-Class Bridge. I

Hiram Leong Fong

On August 24, 1959, Hiram Leong Fong stood at the podium of the U.S. Senate and took the oath of office. It was a great moment in the history of Chinese Americans, for Fong was the first ever to serve in the Congress of the United States. He represented Hawaii, the newest state in the union—the only one with a majority of Asian Americans.

Fong's parents had come to Hawaii from Kwangtung Province in 1872. They worked on a sugar plantation and supported their family of 11 children on their combined wages of $12 a month. Their 7th child, Hiram, born in 1907, followed the classic American success story. As a boy, he supplemented the family income by shining shoes, selling newspapers, and caddying. After graduating from McKinley High School in Honolulu, he worked his way through the University of Hawaii—finding enough time to become editor of the school newspaper as well as a member of the debate team, the volleyball squad, and the rifle team.

Graduating at the beginning of the Great Depression of the 1930s, Fong saved enough money to go to Harvard Law School. He remembered returning home in 1935 "with ten cents in my pocket." After working in Honolulu's city government, he founded the law firm of Fong, Miho, Choy & Robinson, the first multiracial law firm in Hawaii.

Over the next two decades, Fong's shrewd investments made him a multimillionaire. He held interests in shopping centers, a real estate firm, and an insurance company. Even so, when World War II broke out, he signed up to serve in the U.S. Army Air Force, where he rose to the rank of major.

After the war, Fong entered politics, serving 14 years in the territorial legislature. In 1959, when Hawaii became a state, its citizens elected him to one of the two new Senate seats. Though a Republican, Fong received the support of Hawaii's most powerful labor union.

During Fong's three terms as a senator, he toured Asian countries, encouraging understanding and trade between the United States and the emerging Pacific Rim nations. On his trips, he held Hawaii up as an example of a multiracial society that worked for the benefit of all. He retired from politics in 1976.

Two ethnic traditions combine as a young Chinese American in San Antonio, Texas—the oldest Chinese colony in Texas—dresses up as a Mexican American cowboy.

skip the name Liang because it is so difficult for many to pronounce correctly. Even so, the short name has caused much confusion. Some secretaries write it as Keyhole. Others, deciding that such a short last name is impossible, change it arbitrarily to something more common, like Holm or Holt.

When I was sworn as an American citizen, I could have become Keith Ho, or Kenneth Ho, or even Don Ho. I decided to keep my Chinese name; this is one privilege that my new country gives me—the right to maintain my ethnic identity—and I cherish it....

When our daughter was born, we did not give her a Chinese name. We thought that the name should be selected for the child, not for the parents' sake. We would have liked to name her May Hoa Ho, "Ho the Pretty Flower," but imagine the problems that she would face in school among children who like to make fun of "funny" names. So we gave her a "common" American name: Melanie. We hope that she will be as gentle as Melanie Wilkes in "Gone with the Wind." I wonder if Melanie Wilkes' mother ever knew that "Melanie" refers to something black: Would she still have named her so? Only Margaret Mitchell could tell.

So what's in a name? Benjamin Kubelsky changed his name (Jack Benny). Zbigniew Brzezinski did not. I will not either.

Chinese American children often have a sense of belonging to two cultures. They grow up in one yet preserve their roots in another. Sue Jean Lee went to a New York City public school and also attended a Chinese school, which held classes from five to seven o'clock, five days a week.

Chinese school was comprehensive. We learned history, science, social studies (but from a Chinese perspective), geography about China, and in some cases, the world. Very seldom would I mix the lessons from Chinese school with

Among this group of Chinese American children, one wears a shirt with the slogan "Future Presidential Candidate." Outside Hawaii, however, no Chinese Americans have been elected to governorships or the U.S. Congress.

American school. They were two distinct worlds. With only two hours a day, what we did was very selective. We learned composition; we learned how to write [Chinese]. The whole method of teaching was so different from American school. There was a lot of memorization. And penmanship, of course, was using the brush. It was learning how to write Chinese characters, with ink and brush. A very important part of the schooling was poetry.

The Chinese school also had a tradition of a drum and bugle corps. I was a baton twirler. And I was in it for four or five years. All this time that this was going on, it never dawned on me to think about how much out of the mainstream of society we were. It wasn't until I left Chinatown that I realized what a homogenous community Chinatown really was. What a tight, closed environment we lived in....

The [public] school I went to had Italians, blacks, Jews, Puerto Ricans, whites—it was a pretty good mixture. But there was a sense of difference. I got along with people very well in school. I had some very good Italian friends—males and females—but it never occurred to me to date them. Or if I looked for cute guys, I'd be attracted to the Chinese ones and not the Caucasians.... I grew up in a very sheltered environment in Chinatown, and there wasn't the need to look beyond. There were plenty of guys around to date, and enough to have crushes on.

The playwright Frank Chin—a fifth-generation American—has been one of the most assertive and angry of the young Chinese American writers. In Chin's play The Chickencoop Chinaman, *the title character (named Tam Lum) recalls listening to the radio in the 1940s, hoping to find a Chinese counterpart to the popular program "Jack Armstrong, All-American Boy." Tam Lum says:*

I'd spin the dial looking for to hear ANYBODY, CHINESE AMERICAN BOY, ANYBODY, CHINESE AMERICAN BOY anywhere on the dial, doing anything grand on the air, anything at all...I heard of the masked man [the Lone Ranger]. And I listened to him. And in the Sunday funnies he had black hair, and Chinatown was nothin but black hair, and for years, listen, years! I grew blind looking hard through the holes of his funnypaper mask for slanty eyes. Slanty eyes, boys! You see, I knew, children, I knew with all my heart's insight...shhh, listen, children...he wore that mask to hide his Asian eyes!

Yo-Yo Ma

Yo-Yo Ma seemed destined to become a musician. His mother was an opera singer, and his father a violinist, composer, and music teacher. But it was virtually an accident that made Yo-Yo take up the cello, the instrument he has played on concert stages all over the world. When Yo-Yo was four, his older sister began to study the violin. Little brother decided he would surpass her by playing "something bigger"—the cello.

Sitting on a stack of five telephone books, Yo-Yo cradled the instrument between his knees and started his first lessons. His father, a specialist in teaching talented children, guided his son's musical education. Each day, Yo-Yo memorized just two measures of a Bach cello suite. By the time Yo-Yo was five, he knew three Bach pieces by heart.

Yo-Yo Ma's parents had both emigrated from China to Paris, France, where their children were born. The family moved to New York City in 1962, when Yo-Yo was seven. He enrolled in the Juilliard School of Music when he was nine—but by that time he had already performed with Leonard Bernstein and the New York Philharmonic on television and made his New York debut at Carnegie Hall!

Musical prodigies like Yo-Yo Ma sometimes burn out and are not as successful as adults. When Yo-Yo was 17, he showed signs of this tendency while attending a summer music camp. Yo-Yo, in his words, "just went wild." He "never showed up at rehearsals, left my cello out in the rain...midnight escapades to go swimming, and just about everything." At the end of the summer, Yo-Yo left Juilliard to attend college. He was reluctant to take up the rigorous career of a professional musician before finding out if he could do other things as well.

A music teacher at Harvard, Luise Vosgerchian, led him to explore the reasons for his natural gifts of music. She questioned him constantly, making him develop a theory of music that gave him a renewed confidence in his ability. By graduation, Yo-Yo Ma was certain that he wanted a musical career. Since then, he has appeared with symphony orchestras all over the world and is recognized as one of the greatest living cellists.

The one holiday that has always brought Chinese Americans together is the lunar New Year. Traditionally, it lasted two weeks, during which people visited each other to settle debts and wish good luck for the coming year.

CELEBRATIONS

The beginning of the New Year, which in the Chinese lunar calendar falls either in late January or February, is the greatest of Chinese holidays. Louise Leung Larson, who grew up in Los Angeles early in the twentieth century, recalled how her family celebrated.

Papa would start preparations several months in advance by planting Chinese lily bulbs in blue planters. There were at least a dozen or more, so they could be placed in various rooms of the house. They always seemed to bloom just in time for the New Year, and their heady, sweet fragrance filled the house. These were the only flowers he personally planted, and he was very proud of them. Mama would get out the special New Year tablecloth, a red silk embroidered cloth with little glass insets which I thought was the most beautiful thing in the world. On it she would place the gold lacquered dish with many compartments, into which she put different sweets, such as candied ginger, coconut strips, lichee nuts, and sweet and sour plums. The traditional New Year dish was *tsai,* an assortment of vegetables such as hair seaweed, bean threads, snow peas, bamboo shoots, mushrooms, cloud ears fungus,

Four girls holding drums lead a Chinese New Year parade as firecrackers explode around them.

and many others. It required a great deal of chopping, cutting, and cooking, but the result was deliciously flavorful....

At [our house], there was always a constant stream of visitors. Lillie [Louise's older sister] and I would dutifully say *Gong hay fat choy* and the guests would put *lai see* (money wrapped in red paper) in a large straw tray. By the end of the day, the tray would be heaped with wrapped money, not only with coins but with dollar bills. Lillie and I would divide up the money, then Mama would take charge of it, saying it would be put in the bank for us....

We also celebrated the Dragon Boat Festival on the fifth day of the fifth month, and the Autumn Festival during the full moon of the equinox at the end of September. We were never told the origin of these holidays, but they were times of feasting. There were no boats connected with our Dragon Boat Festival, but we had *joong,* a kind of Chinese tamale, made with *naw mai* (sweet rice), *lop cherng* (Chinese sausage), chestnuts, salted duck eggs, *hah mai* (dried shrimp), mushrooms, and other choice ingredients. They were wrapped in ti leaves and boiled....At the Autumn Festival, we ate round moon cakes, some sweet, and some filled with black *do sa.*

Jade Snow Wong recalled the week-long festivities that followed Chinese New Year's Day.

There were Lion Dances daily on the streets.... It was the custom in San Francisco for the Chinese hospital to raise its yearly funds by engaging a "lion" to dance for his money. A group of acrobats...used a large and ferocious-looking but very colorful "lion's head," fitted with bright eyes on springs, and a jaw on hinges. From this head there hung a fancy satin "body" and "tail" piece, sewn together with different-colored scalloped strips of coral, turquoise, red, green, and blue silk. One man who set the tempo for the dance manipulated the head, holding it up in both hands, with only his brightly trousered and slippered legs showing below. As the huge Chinese drums beat in quickening tempos, he danced hard, raised the head high, and jerked it from side to side in an inquiring and delighted manner. His partner, holding up the tail, danced in accompaniment. Their lively movements simulated the stalking, attack, and retreat of a lion.

Citizens of Chinatown co-operated by hanging red paper tied with currency and lettuce leaves in front of their doorways. The lion approached and danced up to the prize. Sometimes, he had to dance onto a stool to reach it. As he stretched his hand out through the mouth to grab the money, his feet keeping time on the stool all the while, the occupant of the house or store threw out strings of burning firecrackers, both to welcome him and to scare away the evil spirits.

As Others Saw Them

During the New Year's festivities around 1900, the Chinese of one California town held an open house and allowed visitors to enter the Chee Kong Tong temple. A newspaper reporter described the scene:

"[I]n the center of the altar was an alcove for the picture of the gods, a group of several...the picture was a couple of feet back from the frame of the alcove of green with carved letters, with a touch of an oriental finish. Hanging from the center of the shrine was the ever burning light held in a brass holder. A pewter holder in front of the picture was filled with burning incense, punks or red candles made of grease. On each side of the alcove were tall, pewter holders for large decorated red candles and the tall punks.

A couple of feet in front was a table-like altar. On this was a pewter bowl for the burning of the fragrant smelling sandal wood by the worshiper. The altar cloth hanging along the front was of brightly embroidered red silk on which were many circular mirrors.

Along the front was a piece of white matting on which the worshiper knelt and in his hands had the incenses and candles which were placed in the bowl. He would pour out libations of wine and burn paper representing the next world, money and clothes."

The Ching Ming festival, which marks the beginning of spring, is a time when families go to the cemetery to pay honor to the dead. The graves are cleaned and decorated, incense is burned, and food and drink are set out for the spirits of ancestors.

Collin Dong was born in 1901, the second son of Dong Teng Seng, who established a successful restaurant in Brooklyn, a Chinese community on the Monterey peninsula of California. Dong recalled that at New Year's:

Mother would dress us in colorful costumes, and along with Father, who was dressed in his best Chinese silken gown, we went from door to door to pay respects to the store owners and to every household in the Chinese communities. The adults of the whole community had prepared little red packages containing 25-cent pieces, called *li shee,* to give to the children. It is an old legend that the giver of *li shee* on New Year's day will have a lucky year. After a day's collection of these red packages, our pockets were usually quite full.

Another traditional Chinese holiday is the Moon Festival. As Jade Snow Wong described it:

According to the Chinese lunar calendar, on the fifteenth day of the eighth month [September] the moon would rise rounder, larger, and more brightly golden than at any other month of the year. Then, specially baked cakes filled with a thick, sweet filling were eaten by the Chinese in recognition of the beautiful, full harvest moon. The round Chinatown moon cakes...were about four inches in diameter and an inch and a half thick. Thin, short, sweet golden pastry was wrapped around rich fillings of ground lotus pods, or candied coconut and melon, or gound sweetened soybean paste. Jade Snow's favorite filling was "five seeds." This was a crunchy, sweet, nutty mixture of lotus pods, almonds, melon seeds, olive seeds, and sesame seeds. Each cake was cut into small wedges, to be enjoyed slowly with tea. Daddy always said that his father in China used to be able to cut his cake into sixteen to thirty-two wedges; one cake would last him all afternoon as he sat on his front porch to eat and drink and leisurely watch the rest of the village go by his door.

Food also plays a role when the Chinese family pays respect to the dead. Sue Jean Lee Suettinger describes how she and her family carry on the ancient traditions.

Whenever I visit my mother during a Chinese holiday and she has the special food and settings out to honor our ancestors, I always light some incense, kowtow [bow] several times and kneel before pictures of my ancestors. In my parents' living room, my dad has [an altar with] pictures of my grandparents and great grandparents. It's set up against the wall, and sometimes oranges and flowers are placed there on special occasions. My parents, who are both in their seventies, still journey twice a year form New Jersey to Washington, D.C., to pay respect to my grandparents at the grave. It's become a way for my family to get together. We would bring a chicken—special ordered because it must have a head and feet—roast pork, and a slice of pork fat and some sweet cakes to the cemetery. Then there are oranges and apples. First we trim the grass around the grave markers, clean them off with water, then we place flowers and the food, and light candles and incense by the graves, offer them the food and then pour three tea cups of Johnny Walker Red on the ground. The food and whiskey is symbolic; it is our way of offering them a special meal each year to honor and remember them, so they won't go hungry. Then we burn paper money, lots of it—gold-sheeted, and even fake bills, so they will have money to spend. Then there are color sheets of paper which symbolize clothing. We burn that too. It's all done in a metal container. And we bring along a cassette of Chinese music which we play, so that my grandparents can enjoy the music while they are eating. We do this every year, and it is a part of what my family feels is our obligation to our ancestors.

Yung Sun Tom told her son Raymond Tom the origins of a favorite family dish.

I learned how to make lots of food from my mother but this recipe, Buddha's Delight, I picked it up on my own while watching my mother and aunts cook it for the whole family from my great-grandmother to my cousin's one-year-old son. I can still remember the condition of the kitchen when I first saw my mother and aunts making it. The whole table was covered with plates of different tasty ingredients. I can still remember my aunts running around, yelling and screaming for the ingredients and my mother putting it into the wok.

It is easy to make because you can use any ingredient you want to, except for meat. We don't use meat because the spirit doesn't like to kill or hurt any kinds of living things. So if we used meat in the dish, it would be an insult to the spirits. It is used to celebrate the birthdays of the spirits on the 1st and 15th of every month on the Chinese calendar. Once it is made, we would first use it to pray to the spirits to pay our respects, then we would eat it. This dish is made mostly to pay tribute to the moon goddess and the warrior of heaven. It is still made today in China to celebrate the spirits' birthdays.

This dish is called Buddha's Delight because monks in the [Buddhist] temples eat it every day because they also don't like to hurt living things. They believed that if they didn't eat meat like the spirit, their souls would be kept clean and their bodies preserved and would remain fit and healthy.

When we first came to America, the first dish I made here was Buddha's Delight because we wanted to thank the spirits for getting us to the new world safely. The boat trip was about one week long and I got seasick about ten times during the trip but I was still well enough to walk off the boat.

To me it doesn't take a special occasion to make it because when I feel like eating it, I will make it which is most of the time. But now I don't make it except for celebrating the spirit's birthday because my family likes eating dishes which have meat in them.

Yung Sun Tom's Buddha's Delight

INGREDIENTS

1/2 cup raw peanuts
7 dried mushrooms
4 oz. dried bean curd (sliced)
1/5 cup dried vegetables
1/5 lb. bean thread
1 lb. soybeans
1 lb. Chinese cabbage
1 wet bean curd
1/4 cup brown sugar
2 cups water
1/2 teaspoon salt

1. Place the peanuts in a pot of boiling water until they are tender.
2. Soak the dried bean curd, mushrooms, dried vegetables, and bean thread in warm water until they are soft.
3. Take the pot off the stove and drain the water from the peanuts.
4. Cut the mushrooms into even slices and chop the soybeans.
5. Put the soybeans into a wok with a little oil.
6. Brown lightly on both sides while adding salt lightly, then take the soybeans out of the wok.
7. Sauté Chinese cabbage with salt.
8. Add mushrooms, peanuts, dried bean curd, dried vegetables, bean threads and soybeans on top, then add water.
9. Cook for 15 minutes.
10. Mix wet bean curd and brown sugar together in a bowl and then place it into the wok.
11. Mix it together and let it heat up for another 5 minutes.
12. Turn off the flame and put it on a plate.

Serves 8 to 10 people as a main course with rice.

THE LEE FAMILY

Members of the third and fourth generation of the Lee family. From left, Stephen Boon, Jr., Eva Lee Lo, Rose Lee Boon, and Sandra Lee Kawano.

Walk along the crowded, busy sidewalks of New York's Chinatown today and you see wares of all kinds displayed—fresh fish and Chinese vegetables, packaged noodles, clothing, toys, watches, jewelry, calculators, magazines and newspapers. Many restaurants cater to those who love good food—from small carry-outs to fancy places where you can order a banquet.

In the historic part of Chinatown, most of the buildings are quite old. Turn off Mott Street onto Pell Street, however, and you see a building with a new stone front. A small brass plate reads Harold L. Lee & Sons. Inside is a modern office. Only a fading orange rice-paper Chinese sign, proudly framed on the wall, gives a clue that this business has been in existence, at this same location, for more than a century.

We sit down with Stephen Boon, Jr., and Sandra Lee Kawano, members of the fourth generation of the Lee family.

Lee Kee Lo, who founded the Lee family business in New York City during the late 1880s.

The village of Toishan, China, where Lee Kee Lo was born.

Q: Who started the family business?

Sandra:

Lee Kee Lo, our great-grandfather, came to the United States as a merchant in 1888. He ended up in New York Chinatown. There were only two or three streets at that time—Mott Street, Pell Street, and the little crooked street named Doyers Street. When he saw 31 Pell Street, he thought it was a propitious, or "lucky," building and rented the storefront. He started the Tai Lung Company with his two brothers. The name means "Great Prosperity." It began as a grocery store and expanded to a curio shop and a meat market.

Lee Kee Lo went back and forth to China to visit his family. When his son, my grandfather Harold, was twelve years old, Lee Kee Lo wanted to bring him to the U.S. He sent him to a program sponsored by the Chinese government at the Mount Hermon School in Northfield, Massachusetts.

Great-grandfather sent him cold, on a ship, without knowing English. This was about 1905. He went to school there for a year, learning English. His name in Chinese is Lee Lun but he took the English name Harold. The company's name today comes from him—Harold L. Lee & Sons. He went back to the village in China after that one year, but learning English gave him an advantage. The village, called Toishan, was in Kwangtung Province, where most of the Chinese people in America in those days came from.

The orange rice-paper sign, more than a century old, that Lee Kee Lo used for his shop.

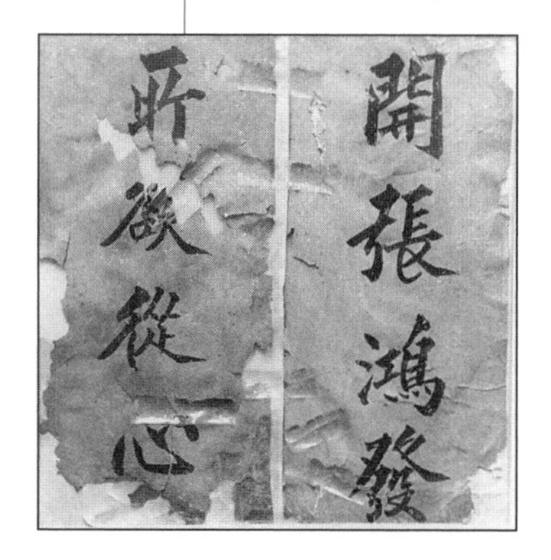

There, in China, he met my grandmother. Her name was Sue Sang Fung. Grandmother came from another village, Kow

Kuang, which just means "Nine Village." They got married in China, but then he brought her over here in 1911 and they lived upstairs on the third floor. We both remember grandmother Sue Sang very well. She and grandfather had six children who lived to adulthood. They all grew up here in the building. It was owned by an Italian landlord, because at that time, Chinese were not allowed to own property.

Harold and Sue Sang's first child, who is Steve's mother, was born in 1913. Her name is Rose Lee, but she had a Chinese name, Lee Sui King, and we all called her Kingy. After Rose there was a second daughter, Eva, and then the first son, my father, Andrew. Later, three other children were born—Martha, Catherine, and Henry.

At this time, grandfather Harold was still carrying on his father's businesses. Then in the 1930s he started a foreign exchange business on the second floor. It was like a bank. In those days, Chinese didn't use the regular banks here, because they were suspicious and there was the language difficulty. If somebody wanted to send money back to China or borrow money, they couldn't do it. So grandfather's business would exchange American currency for Chinese to send money home to China. He also loaned money to people for "ship money" to return home. Laundrymen came in to get bonds for their businesses.

Lee Kee Lo's son Harold L. Lee, as a schoolboy in the United States around 1905.

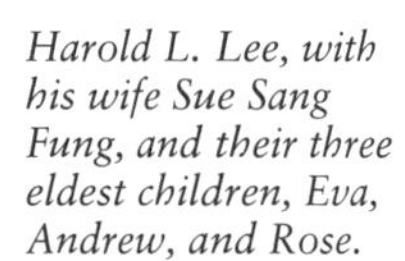

Harold L. Lee, with his wife Sue Sang Fung, and their three eldest children, Eva, Andrew, and Rose.

Q: When did the next generation come into the business?

Sandra:

Steve's mother, Rose, the oldest daughter, started working in the foreign-exchange business when she was about 18. After she married, her husband, Stephen Boon, Sr., joined the business, too. It used to pay people's telephone and electric bills, and perform all sorts of financial services for them. It was a community service. People brought in letters in English to have us read them.

The third generation of Lees in America: Catherine, Eva, Henry, Rose, Martha, and Andrew.

Q: How did the family get into the insurance business?

Sandra:

My father, Andrew, was the first Lee son so that carried a lot of responsibility. After he graduated from high school, grandfather sent him to Shanghai's St. John's University [in China], so he would be fluent in both languages, which would be useful in doing business here. Then Andrew came back to the States and went to Columbia School of Business and got an M.B.A. There he learned about insurance and thought this was a great opportunity. To our Chinese community this was a foreign concept. So my father was a pioneer and became Chinatown's first insurance broker in the late 1930s.

When World War II came, he went into the army. He was in the Central Intelligence Corps as a lieutenant. His sisters kept the business going while he was away.

A silhouette made of Harold L. Lee as a souvenir of the Chicago World's Fair of 1933-34.

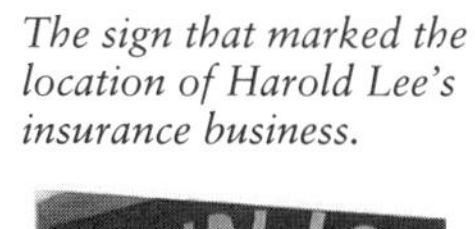

The sign that marked the location of Harold Lee's insurance business.

Eva, as a young girl, took part in this anti-Japanese rally in New York Chinatown in 1928. She can be seen at center right.

Q: Tell us about your mother.

Sandra:

Her name is Shirley Chin. She and my father, Andrew, were married in 1942. My great-grandfather on my mother's side came to the United States and worked on the railroad for a while, according to the family history. But he was captured by Indians. They saw he wasn't white and thought he looked Indian. He was let go after living peacefully with them for two years and went to Baltimore, and then returned to China to marry my great-grandmother.

My grandfather on my mother's side was a paper son. We always knew him by his real name, which was Chin. We didn't know his paper name, but when he had to go to the hospital we had to find his papers, and saw the name Hong Yang. We said, who is Hong Yang? That was his paper name. He immigrated here under that name, but he used his real name for business. There are still a lot of paper sons.

Q: Could you tell us about growing up, your relationship with your family?

Steve:

When my grandfather became successful, he moved the family to a bigger apartment uptown on 14th Street, right off Third Avenue. I grew up there. It was a gathering place for the whole family on holidays. I still remember my grandfather's 70th birthday, which was at the famous old Port Arthur restaurant. It was a huge, festive party because it is traditional in Chinese culture that at certain junctures you will have a large party. There were speeches and celebrations. It made an impression on me because that was one of the main banquet halls which still had all of the cultural tradition of old-world Chinatown. It's no longer there, unfortunately. That's one of the things I remember vividly.

Andrew Lee with one of his grandchildren, Thomas Lee Kawano.

Sandra:

Well, growing up we were always around Chinatown even though we didn't actually live there. There are eleven of us, all cousins, in the fourth generation, but we're all very close, like brothers and sisters. We lived in Flushing, Queens [another part of New York City that has a large population of middle-class Chinese Americans] and then moved out to Long Island. We went to Sunday school at our Lutheran church in Chinatown. It's the True Light Church, where many of the missionaries who sponsored Chinese families worked. So we really saw Chinatown as a gathering place for us, where all the Chinese families got together.

The Lee family gathers at their summer home in Bradley Beach, New Jersey, in 1953. Harold and Sue Sang are at the center of the second row. Eva and Rose stand behind Harold. Sandra and Stephen are in front of Sue Sang.

Q. Why did you join the family business?

Steve:

After attending Long Island University, I went to work for Alexander's Department Stores. I was there for about nine years, worked as department manager up to a position as buyer. I did a lot of international travel as a buyer, going to Asia, Europe, South America. It was very grueling, and I was considering a career change. At that point Sandra's father talked to me about helping to continue the family business. I took him up on the offer and joined the firm around 1979 and have been here ever since. Sandra also joined and the two of us have operated the business since 1986.

Sandra Lee Kawano speaks at a rally in New York. At the far right is the head of the CCBA, the unofficial "mayor" of Chinatown.

Sandra:

My father was going to retire and wanted someone to help him cut back on the work. Steve was already in the business, and I was interested. Because our roots are here in Chinatown, it's a very special place. I think everybody in our family has that tradition of helping others. New immigrants will come into the office because they have received a letter in English and they view our office as a place where they can come for help. Our grandfather started the tradition when he had the foreign exchange here. And we're trying to instill that tradition in our children now. They have an awareness of community, and helping others, of political and social consciousness. I think that is really a family trait of the Lees. We've lasted in this community because we've been so much a part of it. We gained from it and give back. That's our staying power.

Sandra displays the abacus that was a gift from her grandfather Harold. In the picture below, the abacus is set to the number 1888—the year of the founding of the Lee family business.

Q: Do you think your children will keep that? Keep in touch with their roots?

Sandra:

I think so. We're members of the Lee Family Association, whose headquarters are right on the corner. Of course we celebrate Chinese New Year's. There's also a spring rite in which everybody visits the cemetery. Ching Ming. Everybody goes. We see take-out containers of food on the graves, and people burn incense. But we celebrate all holidays. Yesterday, the whole family gathered for Easter.

Asian families spend a lot of time together. Just this last month, we went to California to look at colleges for my older son. They appreciate that we went with them instead of just sending them out by themselves. You show them by example that you want to spend the time together and are concerned about what happens to them. You get that back. It's a two-way street.

Steve's wife is Irish American. Their three sons are Christopher, Jonathan, and Timothy. My husband is Japanese American, third generation, so my children are Asian. They're experiencing both cultures and it's interesting how we're all holding on to bits and pieces of their backgrounds. When our two boys, Thomas Lee and Mark Lee Kawano, were younger we had them enter a Boy Scout troop in Chinatown because we thought that was important. That was the same troop my father was in and it was a link with the generations.

Members of the fifth generation of the Lee family, among them the grandchildren of Andrew, Rose, and Eva, appear in this family portrait, taken at Bradley Beach, New Jersey, in 1981.

CHINESE AMERICAN TIMELINE

1830s
Chinese sugarcane workers go to Hawaii; Chinese sailors arrive in New York City.

1849
Discovery of gold in California brings Chinese in large numbers to the United States for the first time.

1851
Chinese in San Francisco form Sze Yup and Sam Yup district associations.

1865
Central Pacific Railroad recruits first Chinese laborers.

1868
United States and China sign Burlingame Treaty, recognizing right of citizens to enter each country.

1869
2,000 Chinese complete western part of transcontinental railroad.

1871
Anti-Chinese violence in Los Angeles.

1878
U.S. Supreme Court denies right of Chinese to become naturalized U.S. citizens.

1880
District associations in San Francisco form Chung Wah Kung Saw, Chinese Consolidated Benevolent Association, also known as Six Companies.

1882
Chinese Exclusion Act prohibits Chinese from entering the United States; merchants, diplomats, and students are excepted.

1885
Massacre of Chinese miners at Rock Springs, Wyoming.

1888
Lue Gim Gong develops orange that becomes basis of Florida fruit-growing industry.

1894
In Honolulu, Sun Yat-sen founds group dedicated to overthrow of Manchu dynasty.

1898
U.S. Supreme Court rules that children of Chinese descent born in the United States are legally citizens.

1906
San Francisco earthquake levels Chinatown: destruction of records allows Chinese living there to claim they were born in the United States.

1909
Angel Island Immigration Station established in San Francisco to screen Chinese immigrants claiming to be relatives of Chinese American citizens.

1911
Chinese American men cut off queues after overthrow of the Manchu dynasty in China.

1924
Immigration Act prohibits Chinese American citizens from bringing wives and children to the United States.

1937
Japan invades China; Chinese Americans stage demonstrations against American trade with Japan.

1941
United States declares war on Japan, allying itself with China in World War II.

1943
Chinese Exclusion Act repealed.

1949
Mao Zedong establishes communist government in China; Nationalists set up independent regime on Taiwan.

1959
Hawaii becomes 50th state; first Chinese American, Hiram Leong Fong, elected to U.S. Senate.

1965
Immigration law abolishes "national origins" as basis for immigrant quotas. Chinese permitted to enter on equal basis with other immigrants.

1980s
Chinese American population exceeds 1 million for the first time.

FURTHER READING

General Accounts of Chinese American History

Chen, Jack. *The Chinese of America.* San Francisco: Harper & Row, 1980.

Melendy, H. Brett. *The Oriental Americans.* New York: Hippocrene Books, 1972.

Steiner, Stan. *Fusang: The Chinese Who Built America.* New York: Harper Colophon Books, 1979.

Sung, Betty Lee. *Mountain of Gold: The Story of the Chinese in America.* New York: Macmillan, 1967.

Takaki, Ronald. *Strangers from a Different Shore.* New York: Penguin Books, 1989.

Tsai, Shih-Shan Henry. *China and the Overseas Chinese in the United States, 1868–1911.* Fayetteville: University of Arkansas Press, 1983.

Specific Aspects of Chinese American History

Dicker, Laverne Mau. *The Chinese in San Francisco: A Pictorial History.* New York: Dover Publications, 1979.

Kinkead, Gwen. *Chinatown: Portrait of a Closed Society.* New York: Harper Collins, 1992.

Kwong, Peter. *The New Chinatown.* New York: Noonday Press, 1987.

Lydon, Sandy. *Chinese Gold: The Chinese in the Monterey Bay Region.* Capitola, Calif.: Capitola Book Co., 1985.

Siu, Paul C.H. *The Chinese Laundryman.* New York: New York University Press, 1987.

Wu, Cheng-Tsu, ed. *Chink! A Documentary History of Anti-Chinese Prejudice in America.* New York: World, 1972.

Yung, Judy. *Chinese Women of America: A Pictorial History.* Seattle: University of Washington Press, 1986.

First-Person Accounts of Chinese American Life

Char, Tin-Yuke. *The Sandalwood Mountains.* Honolulu: University Press of Hawaii, 1975.

Holt, Hamilton, ed. *The Life Stories of Undistinguished Americans as Told by Themselves.* 1906. Reprint. New York: Routledge, 1990.

Huang, Joe, and Sharon Quan Wong, eds. *Chinese Americans: Myths and Realities.* San Francisco: Association of Chinese Teachers, 1977.

Larson, Louise Leung. *Sweet Bamboo: Saga of a Chinese American Family.* Los Angeles: Chinese Historical Society of America, 1989.

Lee, Joann Faung Jean. *Asian American Experiences in the United States.* Jefferson, N.C.: McFarland, 1991.

Lord, Bette Bao. *Legacies.* New York: Knopf, 1990.

Lowe, Pardee. *Father and Glorious Descendant.* Boston: Little, Brown, 1943.

Nee, Victor G., and Brett de Bary Nee. *Longtime Californ'.* Stanford, Calif.: Stanford University Press, 1986.

Wang, An. *Lessons.* Reading, Mass.: Addison-Wesley, 1986.

Wong, Jade Snow. *Fifth Chinese Daughter.* New York: Harper & Row, 1950.

Yep, Laurence. *The Lost Garden.* Englewood Cliffs, N.J.: Julian Messner, 1991.

Poems, Stories, Novels, and Plays by Chinese Americans

Chin, Frank. *The Chickencoop Chinaman and The Year of the Dragon.* Seattle: University of Washington Press, 1981.

———. *Donald Duk.* Minneapolis: Coffee House Press, 1991.

Chin, Frank, Jeffry Paul Chan, Lawrence Fusao Inada, and Shawn Wong, eds. *Aiiieeeee!: An Anthology of Asian-American Writers.* Washington, D.C.: Howard University Press, 1974.

Chu, Louis. *Eat a Bowl of Tea.* New York: Lyle Stuart, 1961.

Hom, Marlon K. *Songs of Gold Mountain: Cantonese Rhymes from San Francisco Chinatown.* Berkeley: University of California Press, 1987.

Kingston, Maxine Hong. *China Men.* New York: Ballantine, 1981.

———. *The Woman Warrior.* New York: Knopf, 1976.

Lai, Him Mark, Genny Lim, and Judy Yung. *Island: Poetry and History of Chinese Immigrants on Angel Island, 1910–1940.* Seattle: University of Washington Press, 1980.

Lee, C. Y. *The Land of the Golden Mountains.* New York: Meredith Press, 1967.

Lee, Gus. *China Boy.* New York: Dutton, 1991.

Lo, Steven C. *The Incorporation of Eric Chung.* Chapel Hill, N.C.: Algonquin Books of Chapel Hill, 1989.

Tan, Amy. *The Joy Luck Club.* New York: Putnam's, 1989.

———. *The Kitchen God's Wife.* New York: Putnam's, 1991.

Yep, Laurence. *The Star Fisher.* New York: Morrow, 1991.

TEXT CREDITS

Main text

p. 12, top: Huie Kin, *Reminiscences* (Peiping, China: San Yu Press, 1932; New York: G.L. Trigg, 1982), 6.

p. 12, bottom: Hamilton Holt, ed., *The Life Stories of Undistinguished Americans.* Reprint. (1906; reprint, New York: Routledge, 1990), 174–78.

p. 13: Louise Leung Larson, *Sweet Bamboo* (Los Angeles: Chinese Historical Society of Southern California, 1990), 3–5.

p. 14: Chung Kun-ai, *My Seventy-Nine Years in Hawaii (1879–1958)* (Hong Kong: Cosmorama Pictorial Publisher, 1960), 1–29.

p. 16: Elizabeth Wong, "Leaves from the Life Story of a Chinese Immigrant," *Social Process in Hawaii* 2 (1936): 39–42.

p. 17: An Wang, *Lessons* (Reading, Mass.: Addison-Wesley, 1986), 13.

p. 18: Kin, *Reminiscences,* 17–19.

p. 19: Holt, *The Life Stories,* 178–79.

p. 20, top: Joann Faung Jean Lee, ed., *Asian American Experiences in the United States* (Jefferson, N.C.: McFarland, 1991), 79.

p. 20, bottom: Tricia Knoll, *Becoming Americans.* (Portland, Oreg.: Coast to Coast Books, 1982).

p. 26: Kin, *Reminiscences,* 17–20.

p. 27: Chung, *My Seventy-Nine Years,* 1–29.

p. 28, top: Kin, *Reminiscences,* 20–21.

p. 28, bottom: Elizabeth Wong, "Leaves from the Life Story," 39–42.

p. 29: From *Longtime Californ'* by Victor G. Nee and Brett de Bary Nee. Copyright © 1972, 1973 by Victor G. and Brett de Bary Nee. Reprinted by permission of Pantheon Books, a division of Random House, Inc.

p. 30: Kin, *Reminiscences,* 20–23.

p. 31, top: Chung, *My Seventy-Nine Years,* 1–29.

p. 31, bottom: Holt, *The Life Stories,* 179.

p. 32, top: Elizabeth Wong, "Leaves from the Life Story," 39–42.

p. 32, bottom: Quan Ngo, Cleveland High School, Seattle, Wash., 1980; quoted in Knoll, *Becoming Americans,* 267–70.

p. 38: Kin, *Reminiscences,* 24–25.

p. 39, top: From *Longtime Californ'* by Victor G. Nee and Brett de Bary Nee. Copyright © 1972, 1973 by Victor G. and Brett de Bary Nee. Reprinted by permission of Pantheon Books, a division of Random House, Inc.

p. 39, bottom: Copyright © 1989 by Steven C. Lo. Reprinted by permission of Algonquin Books of Chapel Hill, a division of Workman Publishing Company, New York, N.Y.

p. 40: Genny Lim and Judy Yung, "Our Parents Never Told Us," *California Living Magazine,* Jan. 23, 1977.

p. 41, top: Him Mark Lai, Genny Lim, and Judy Yung, *Island: Poetry and History of Chinese Immigrants on Angel Island* (Seattle: Univ. of Washington Press, 1980), 48. Copyright © 1980 by the HOC DOI (History of Chinese Detained on Island) Project.

p. 41, bottom: Lai, Lim, and Yung, *Island,* 75. Copyright © 1980 by the HOC DOI (History of Chinese Detained on Island) Project.

pp. 42–43: Lai, Lim, and Yung, *Island,* 54, 58, 66, 84, 124. Copyright © 1980 by the HOC DOI (History of Chinese Detained on Island) Project.

p. 44: Cheng-tsu Wu, *Chink!* (New York: World Publishing, 1972), 97–102.

p. 45: Peter C. Y. Leung, *One Day, One Dollar: Locke, California and the Chinese Farming Experience in the Sacramento Delta* (El Cerrito, Calif.: Chinese/Chinese American History Project), 48.

p. 46, top: From *Longtime Californ'* by Victor G. Nee and Brett de Bary Nee. Copyright © 1972, 1973 by Victor G. and Brett de Bary Nee. Reprinted by permission of Pantheon Books, a division of Random House, Inc.

p. 46, bottom: Pardee Lowe, *Father and Glorious Descendant* (Boston: Little, Brown, 1943), 122.

p. 47: Lee, *Asian American Experiences,* 5.

p. 52, top: From *China Men* by Maxine Hong Kingston. Copyright © 1980 by Maxine Hong Kingston. Reprinted by permission of Alfred A. Knopf, Inc.

p. 52, bottom: C. Y. Lee, *Land of the Golden Mountain* (New York: Meredith Press, 1967), 103–5.

p. 54, top: "Chinese Fisheries in California," *Chamber's Journal* 1 (1854): 48.

p. 54, bottom: Keith Wheeler, *The Alaskans* (Alexandria, Va.: Time-Life Books, 1977), 218.

p. 56, top: Henryk Sienkiewicz, "The Chinese in California," *California Historical Society Quarterly* 34 (1955): 301–16.

p. 56, bottom: From *Longtime Californ'* by Victor G. Nee and Brett de Bary Nee. Copyright © 1972, 1973 by Victor G. and Brett de Bary Nee. Reprinted by permission of Pantheon Books, a division of Random House, Inc.

p. 58: From *China Men* by Maxine Hong Kingston. Copyright © 1980 by Maxine Hong Kingston. Reprinted by permission of Alfred A. Knopf, Inc.

p. 62, top: From *Longtime Californ'* by Victor G. Nee and Brett de Bary Nee. Copyright © 1972, 1973 by Victor G. and Brett de Bary Nee. Reprinted by permission of Pantheon Books, a division of Random House, Inc.

p. 62, bottom: information from Cheng-tsu Wu, *Chink!,* and E. C. Sandmeyer, *The Anti-Chinese Movement in California* (Urbana: University of Illinois Press, 1939).

p. 63: *Sacramento Union,* Jan. 23, 1855, p. 2, quoted in H. Brett Melendy, *The Oriental Americans* (New York: Hippocrene Books, 1972).

p. 64: Kin, *Reminiscences,* 26–27.

p. 65, top: *San Francisco Argonaut,* Aug. 10, 1878.

p. 65, bottom: Sandy Lydon, *Chinese Gold: The Chinese in the Monterey Bay Region* (Capitola, Calif.: Capitola Book Company, 1985), 134–35.

p. 66: Cheng-tsu Wu, *Chink!,* 152–55.

p. 72: quote in introduction from *Alta California,* Aug. 22, 1851.

p. 72: Kin, *Reminiscences,* 25.

p. 73: quotes in introduction from Connie Young Yu, "From Tents to Federal Projects: Chinatown's Housing History," in *The Chinese American Experience: Papers from the Second National Conference on Chinese American Studies* (San Francisco: Chinese Historical Society of America and the Chinese Cultural Foundation, 1984).

p. 73: From *Longtime Californ'* by Victor G. Nee and Brett de Bary Nee. Copyright © 1972, 1973 by Victor G. and Brett de Bary Nee. Reprinted by permission of Pantheon Books, a division of Random House, Inc.

p. 75, top: Ching-chao Wu, "Chinatowns: A Study in Symbiosis and Assimilation," Ph.D. dissertation, University of Chicago, 1928, quoted in Roger Daniels, *Asian America* (Seattle: University of Washington Press, 1988), 70.

p.75, bottom: Amy Tan, *The Kitchen God's Wife.* (New York: Putnam, 1991), 18–19.

p. 76: both from "Remembering New York Chinatown," Chinatown History Museum, New York City, 1992.

p. 78: B. E. Lloyd, *Lights and Shades of San Francisco,* quoted in Yu, "From Tents to Federal Projects."

p. 79, top: Louis Chu, *Eat A Bowl of Tea* (New York: Lyle Stuart, 1961).

p. 79, middle: Ronald Takaki, *Strangers from a Different Shore* (New York: Penguin Books, 1989), 232.

p. 79, bottom: Betty Lee Sung, *Mountain of Gold: The Story of the Chinese in America* (New York: Macmillan, 1967), 140–42.

p. 81 (both): Paul C. H. Siu, *The Chinese Laundryman: A Study in Isolation* (New York: New York University Press, 1987), 157–58.

p. 82: From *Longtime Californ'* by Victor G. Nee and Brett de Bary Nee. Copyright © 1972, 1973 by Victor G. and Brett de Bary Nee. Reprinted by permission of Pantheon Books, a division of Random House, Inc.

p. 83, top: From *Longtime Californ'* by Victor G. Nee and Brett de Bary Nee. Copyright © 1972, 1973 by Victor G. and Brett de Bary Nee. Reprinted by permission of Pantheon Books, a division of Random House, Inc.

p. 83, middle: Reprinted from *American Mosaic: The Immigrant Experience in the Words of Those Who Lived It,* by Joan Morrison and Charlotte Fox Zabusky, by permission of the University of Pittsburgh Press. © 1980, 1993 by Joan Morrison and Charlotte Fox Zabusky.

p. 83, bottom: William Hoy, *The Chinese Six Companies.* (San Francisco: Chinese Consolidated Benevolent Association, 1942), 3.

p. 85, top: From *Longtime Californ'* by Victor G. Nee and Brett de Bary Nee. Copyright © 1972, 1973 by Victor G. and Brett de Bary Nee. Reprinted by permission of Pantheon Books, a division of Random House, Inc.

p. 85, bottom: Eng Ying Gong and Bruce Grant, *Tong Wars* (New York: Nicholas L. Brown, 1930), 10.

p. 86: From *Asian American Experiences in the United States: Oral Histories of First to Fourth Generation Americans from China, the Philippines, Japan, India, the Pacific Islands, Vietnam and Cambodia.* © 1991 Joann Faung Jean Lee, McFarland & Company, Inc., Publishers, Jefferson, N.C. 28640.

p. 87: "A Chinese Family in Hawaii," *Social Process in Hawaii* 3 (1937): 50–55.

p. 89, top: "A Chinese Family in Hawaii," 50–55.

p. 89, bottom: Sung, *Mountain of Gold,* 151–52.

p. 91: Elizabeth Wong, "Leaves from the Life Story," 39–42.

p. 92: Siu, *The Chinese Laundryman,* 72–74.

p. 93: Siu, *The Chinese Laundryman,* 116–17.

p. 94, top: Siu, *The Chinese Laundryman,* 130.

p. 94, bottom: New York Chinatown History Project.

p. 95, top: Stephen Williams, "The Chinese in the California Mines," Ph.D. dissertation, Stanford University, 1930, p. 59; quoted in Knoll, *Becoming Americans,* 19–20.

p. 95, bottom: From *Longtime Californ'* by Victor G. Nee and Brett de Bary Nee. Copyright © 1972, 1973 by Victor G. and Brett de Bary Nee. Reprinted by permission of Pantheon Books, a division of Random House, Inc.

p. 96: Tommy Woo, "12 Hours a Day," *Sojourner IV,* Asian Writers Project, Berkeley Unified School District, 1974, quoted in Knoll, *Becoming Americans,* 40.

p. 98: From *Longtime Californ'* by Victor G. Nee and Brett de Bary Nee. Copyright © 1972, 1973 by Victor G. and Brett de Bary Nee. Reprinted by permission of Pantheon Books, a division of Random House, Inc.

p. 99: Takaki, *Strangers from a Different Shore,* 371.

p. 100, top: Takaki, *Strangers from a Different Shore,* 372.

p. 100, bottom: Takaki, *Strangers from a Different Shore,* 373.

p. 106, top: From *Longtime Californ'* by Victor G. Nee and Brett de Bary Nee. Copyright © 1972, 1973 by Victor G. and Brett de Bary Nee. Reprinted by permission of Pantheon Books, a division of Random House, Inc.

p. 106, bottom: Anne Lee, "Just a Trap," *Sojourner IV,* quoted in Knoll, *Becoming Americans,* 273–74.

p. 108: Reprinted from *American Mosaic: The Immigrant Experience in the Words of Those Who Lived It,* by Joan Morrison and Charlotte Fox Zabusky, by permission of the University of Pittsburgh Press. © 1980, 1993 by Joan Morrison and Charlotte Fox Zabusky.

p. 109, top: Joann Faung Jean Lee, *Asian American Experiences,* 78–81.

p. 109, bottom: Gwen Kinkead, *Chinatown: A Portrait of a Closed Society* (New York: HarperCollins, 1992), 25.

p. 110: Copyright © 1989 by Steven C. Lo. Reprinted by permission of Algonquin Books of Chapel Hill, a division of Workman Publishing Company, New York, N.Y.

p. 111: Knoll, *Becoming Americans,* 315–16.

p. 112: Joann Faung Jean Lee, *Asian American Experiences,* 39–41.

p. 113: Frank Chin, *The Chickencoop Chinaman and The Year of the Dragon* (Seattle: University of Washington Press, 1981), 31–32.

p. 114: Larson, *Sweet Bamboo,* 77–79.

p. 115: Jade Snow Wong, *Fifth Chinese Daughter* (New York: Harper & Row, 1950), 41–42.

p. 116, top: Lydon, *Chinese Gold,* 413–14.

p. 116, middle: Jade Snow Wong, *Fifth Chinese Daughter,* 42–43.

p. 116, bottom: Joann Faung Jean Lee, *Asian American Experiences,* 164.

p. 117: "From the Kitchens of the Lower East Side," vol. 2, published by Seward Park High School, New York City, June 1986.

Sidebars

p. 13: Betty Lee Sung, *Mountain of Gold* (New York: Macmillan, 1967), 172.

p. 14: Sung, *Mountain of Gold,* 172.

p. 18: Stan Steiner, *Fusang: The Chinese Who Built America* (New York: Harper Colophon Books, 1979), 113–14.

p. 19: "Remembering New York Chinatown," Chinatown History Museum.

p. 20: Alexander McLeod, *Pigtails and Gold Dust* (Caldwell, Idaho: Caxton Press, 1947), 23.

p. 21: Sung, *Mountain of Gold,* 10.

p. 26: Steiner, *Fusang,* 3.

p. 28: Marlon K. Hom, *Songs of Gold Mountain* (Berkeley: University of California Press, 1987), 160.

p. 29: Tin-yuke Char, *The Sandalwood Mountains* (Honolulu: University Press of Hawaii, 1975), 67.

p. 32: Gwen Kinkead, *Chinatown: A Portrait of a Closed Society* (New York: HarperCollins, 1992), 160–61.

p. 39: Albert S. Evans, "From the Orient Direct," *Atlantic Monthly* 24 (1869): 543–47.

p. 43: Him Mark Lai, Genny Lim, and Judy Yung, *Island: Poetry and History of Chinese Immigrants on Angel Island* (Seattle: University of Washington Press, 1980), 136. Copyright © 1980 by the HOC DOI (History of Chinese Detained on Island) Project.

p. 45: New York Chinatown History Museum, *Bugaoban*, Winter-Spring issue, 1985, page 5.

p. 52: Stephen Williams, "The Chinese in The California Mines," Ph.D. dissertation, Stanford University, 1930; quoted in Tricia Knoll, *Becoming Americans* (Portland, Oreg.: Coast to Coast Books, 1982), 16.

p. 53: Daniel Chu and Samuel Chu, *Passage to the Golden Gate* (Garden City, N.Y.: Zenith Books, 1967), 36.

p. 58: quoted in Steiner, *Fusang,* 133.

p. 59: *Santa Cruz Sentinel,* November 1879, quoted in Sandy Lydon, *California Gold* (Capitola, Calif.: Capitola Book Co., 1985), 78.

p. 60: Jack Chen, *The Chinese of America* (San Francisco: Harper & Row, 1980), 77.

p. 62: Sucheng Chan, *Asian Americans: An Interpretive History* (Boston: Twayne, 1991), 48.

p. 64: James P. Shenton and Gene Brown, eds., *Ethnic Groups in Ameican Life* (New York: Arno Press, 1978), 123.

p. 65: Shih-Shan Henry Tsai, *China and the Overseas Chinese in the United States, 1868–1911* (Fayetteville: University of Arkansas Press, 1983), 108.

p. 77: "Red Boat on the Canal: Cantonese Operatic Arts in New York City Chinatown," Chinatown History Museum.

p. 78: Hom, *Songs of Gold Mountain,* 97

p. 79: Daphne Marlett and Carole Itter, *Opening Doors: Vancouver's East End,* Sound Heritage Series, vol. 8, p. 41; quoted in Knoll, *Becoming Americans,* 16.

p. 83: quote within story from McLeod, *Pigtails and Gold Dust,* 53-54.

p. 87: *Salinas* (California) *Democrat,* Feb. 14, 1891; quoted in Lydon, *California Gold,* 298.

p. 89: "Remembering New York Chinatown," Chinatown History Museum.

p. 90: "Remembering New York Chinatown," Chinatown History Museum.

p. 93: *New York Illustrated News,* June 4, 1853, p. 359; quoted in H. Brett Mellendy, *The Oriental Americans* (New York: Hippocrene Books, 1972), 54.

p. 94: Wells, Fargo History Room, San Fransico.

p. 96: McLeod, *Pigtails and Gold Dust,* 33.

p. 100: Ronald Takaki, *Strangers from a Different Shore* (New York: Penguin Books, 1989), 373.

p. 115: Ernest Otto, *Santa Cruz Sentinel,* July 30, 1944; quoted in Lydon, *California Gold,* 258.

p. 117: adapted from "From the Kitchens of the Lower East Side," vol. 2, published by Seward Park High School, New York City, June 1986.

PICTURE CREDITS

Arizona Historical Society Library: 5 (#5028, Buehman Coll.), 68 (#89203, Buehman Coll.), 92 top (#25164, Reynolds Coll.); Jung Family Photograph, Balch Institute for Ethnic Studies Library; 86 top; Bancroft Library: 48; Bettmann Archive: 17 top, 20, 65, 107 (both), 108, 110 top, 111; Bishop Museum: 55 top, 89; California Historical Society: 30 middle (FN-01002), 38 top (FN-13088, artist: Capt. W. Swasey), 38 bottom (FN-13113), 41 bottom (FN-18240), 53 top (FN-13890, photographer: Eadweard J. Muybridge, 54 bottom left (FN-02324, photographer: Arnold Genthe), 54 bottom right (FN-23112, Eadweard J. Muybridge), 57 top (FN-12616), 74 top (FN-01003, E. N. Sewell), 80 bottom (FN-04899), 83 (FN-21487), 84 top (FN-02352), 93 (FN-29905); California Department of Parks and Recreation, Mak Takahashi: 43; California State Library: 52, 61, 71, 73; Chinatown History Museum, New York: 21 top, 29, 36, 45, 64 right, 74 middle, 76 (both), 77 (all), 85, 88 bottom, 90, 95 top, 98, 99 (both), 101; photo by Paul Calhoun, courtesy of Chinatown History Museum, New York: 94, 106, 116; photo by Robert Glick, courtesy of Chinatown History Museum, New York: 102, 112 bottom; photo by Margaret Yuen, courtesy of Chinatown History Museum, New York: 110 bottom; Connie Chung: 109; Connecticut Historical Society, Hartford, Connecticut: 33 (both); Denver Public Library, Western History Department: 59 top, 60; Hawaii State Archives: 22, 25, 30 top, bottom, 50, 74 bottom, 81 top, 86 bottom; T. Hoobler: 79, 104; Idaho State Historical Society: 56 (#62-44.7); International Creative Management: 113; Dong Kingman: 97; Him Mark Lai: 57 bottom; Corky Lee: 67 right, 105 bottom; courtesy of the Lee Family: 118, 119, 120, 121; Library of Congress: 18, 19, 24, 31, 39, 54 top, 78, 80 top, 81 bottom, 84 bottom, 87, 91, 114 top; courtesy of Raymond L. Lim: cover; courtesy of Bette Bao Lord: 6, 7; Nikola Miller: 114 bottom; National Archives: 34, 37, 41 top, 42, 44, 100 (both); National Archives, Mid-Atlantic Region: 46, 47, 92 bottom; Nevada Historical Society: 53 bottom, 57 middle; Oakland Museum: 4; Oakland Museum, M. Lee Fatheree: 27; courtesy of The Oakland Museum History Department: frontispiece; Ohlinger: 66; Oregon Historical Society: 105 top; Courtesy Peabody and Essex Museum, Salem, Mass.: 8, 10, 11, 12, 14 (both), 15 (both), 16, 17 bottom; Royal Asiatic Society: 13, 21 bottom; San Francisco Maritime National Historic Park, Procter Collection: 55 bottom; San Francisco Office, Records of the Immigration and Naturalization Service, RG 85, National Archives, Pacific Sierra Region, San Bruno, CA: 40 (both), 70; San Francisco Public Library, 72 top; Smithsonian Institution: 63 top, 64 left; Society of California Pioneers: 82 bottom; Southern Pacific Railroad: 58 (both), 59 bottom, 60 top; K. Caugler, United Nations High Commission for Refugees: 32; Courtesy of Wells, Fargo Bank: 26, 72 bottom; Virginia Wong, copy from Institute of Texan Cultures: 112 top (76-449); Wyoming State Museum: 51, 67 left.

INDEX

References to illustrations are indicated by page numbers in *italics*.

ACKNOWLEDGMENTS

Our sincere thanks to Mei-Lin Liu and the staff of the New York Chinatown History Museum, without whose generous help this book would not have been possible.

We also wish to express our gratitude for the assistance and kindness of Patricia Akre of the San Francisco Public Library; R. Joseph Anderson, Monique Bourque and Jennifer Van Vlanderen of the Balch Institute for Ethnic Studies; Shawn Aubitz of the National Archives, Mid-Atlantic Region; Diane Bruce of the Institute of Texan Cultures at San Antonio; Barbara Bush of the Arizona Historical Society; Paul Calhoun; Robert J. Chandler of the Wells, Fargo Bank History Department; Jean J. Charbonnet and Susan Shaner of the Hawaii State Archives; Paula West Chavoya of the Wyoming State Museum; Connie Chung; Phil Earl of the Nevada Historical Society; Marcia Eymann and Drew Johnson of the Oakland Museum; Mike Furtney of the Southern Pacific Company; Robert Glick; Nancy Heywood of the Peabody and Essex Museum; Marge Kemp and Kanani Texeira of the Bishop Museum; Dong Kingman; Him Mark Lai; Evelyn Lane of the Hawaii Chinese Historical Center; Dan Larson and Jane Leung Larson; Corky Lee; Raymond Lim; Jessica Lind of the University of Washington Press; Manni Liu of the Chinese Culture Center of San Francisco; Waverly Lowell of the National Archives, Pacific Sierra Region; Robert MacKimmie and Haley Moffett of the California Historical Society; Nikola Miller; Richard Ogar of the Bancroft Library; Michael Pollock of the Royal Asiatic Society, London; Dr. Gretchen Schneider and Stanleigh Bry of the Society of California Pioneers; Betty Lee Sung; Mikki Tint of the Oregon Historical Society; Margie Turnmire of McFarland and Company; Chris Watkins of the *Santa Cruz* (CA) *Sentinel;* and Virginia Y. Wong.

Thanks and appreciation to our editors Tara Deal and Nancy Toff for all their efforts.

We owe special thanks to Sandra Lee Kawano, Stephen Boon, Jr., Rose Lee Boon, and Eva Lee Lo for sharing their memories and their own Chinese American family album.

ABOUT THE AUTHORS

Dorothy and Thomas Hoobler have published more than 50 books for children and young adults, including *China: History, Culture, Geography; Margaret Mead: A Life in Science; Vietnam: Why We Fought; Showa: The Age of Hirohito;* and *Photographing History: The Career of Mathew Brady*. Their works have been honored by the Society for School Librarians International, the Library of Congress, the New York Public Library, the National Council for Social Studies, and *Best Books for Children*, among other organizations and publications. The Hooblers have also written several volumes of historical fiction for children, including *Frontier Diary, The Summer of Dreams,* and *Treasure in the Stream.* Dorothy Hoobler received her master's degree in American history from New York University and worked as a textbook editor before becoming a full-time freelance editor and writer. Thomas Hoobler received his master's degree in education from Xavier University and he previously worked as a teacher and textbook editor.